PDE Modeling of Tissue Engineering and Regenerative Medicine

PDE Modeling of Tissue Engineering and Regenerative Medicine

Computer Analysis in R

William E. Schiesser

ACADEMIC PRESS

An imprint of Elsevier

Academic Press is an imprint of Elsevier
125 London Wall, London EC2Y 5AS, United Kingdom
525 B Street, Suite 1650, San Diego, CA 92101, United States
50 Hampshire Street, 5th Floor, Cambridge, MA 02139, United States
The Boulevard, Langford Lane, Kidlington, Oxford OX5 1GB, United Kingdom

Notices

Knowledge and best practice in this field are constantly changing. As new research and experience broaden our understanding, changes in research methods, professional practices, or medical treatment may become necessary.

Practitioners and researchers must always rely on their own experience and knowledge in evaluating and using any information, methods, compounds, or experiments described herein. In using such information or methods they should be mindful of their own safety and the safety of others, including parties for whom they have a professional responsibility.

To the fullest extent of the law, neither the Publisher nor the authors, contributors, or editors, assume any liability for any injury and/or damage to persons or property as a matter of products liability, negligence or otherwise, or from any use or operation of any methods, products, instructions, or ideas contained in the material herein.

ISBN: 978-0-443-18740-7

For information on all Academic Press publications
visit our website at https://www.elsevier.com/books-and-journals

Publisher: Mara E. Conner
Acquisitions Editor: Chris Katsaropoulos
Editorial Project Manager: Tom Mearns
Production Project Manager: Nirmala Arumugam
Cover Designer: Matthew Limbert

Typeset by VTeX

Working together
to grow libraries in
developing countries

www.elsevier.com • www.bookaid.org

Contents

Preface

This book is directed to the computer-based modeling of *tissue engineering* and *regenerative medicine* (TERM). As explained in [2],

> *The goal of tissue engineering is to assemble functional constructs that restore, maintain, or improve damaged tissues or whole organs.*

> *Regenerative medicine is a broad field that includes tissue engineering but also incorporates research on self-healing – where the body uses its own systems, sometimes with help foreign biological material to recreate cells and rebuild tissues and organs. The terms "tissue engineering" and "regenerative medicine" have become largely interchangeable, as the field hopes to focus on cures instead of treatments for complex, often chronic, diseases.*

The TERM mathematical model discussed in this book consists of a system of eight partial differential equations (PDEs) with dependent variables u_{1m}, u_{2m}, u_1, u_2, u_3, u_4, u_5, u_6 [1]. These variables are explained in the following table:

normalized variable	physical interpretation
u_{1m}	O_2 level in seeded stem cells
u_{2m}	nutrient level in seeded stem cells
u_1	stem cell density
u_2	transit-amplifying (TA) cell density
u_3	terminally differentiated (TD) cell density
u_4	signaling (regulatory) biomolecule 1 concentration produced by u_1, u_2, u_3, u_6
u_5	signaling (regulatory) biomolecule 2 concentration produced by u_1, u_2, u_3, u_6
u_6	signaling (regulatory) biomolecule 3 concentration produced by u_4, u_5
r	radial distance in stem cells; lower boundary, $r = r_l = 0$; upper boundary at external surface, $r = r_u$
z	axial distance in stem cells; lower boundary, $z = z_l$; upper boundary at scaffold surface, $z = z_u$
t	time

The eight PDE model is coded (programmed) as a set of routines in R, a quality, open-source, scientific programming system [3]. r, z, t are the PDE independent variables. r, z are two spatial variables in cylindrical coordinates (r, θ, z) reflecting the geometry of the scaffold and seeded stem cells of the tissue construct (TC), and t is time. These independent variables therefore define the spatiotemporal basis of the PDE model. $u_3(r, z, t)$ is of particular interest since it expresses the level of the terminally differentiated (TD) cell density, the final goal/product of tissue engineering.

The TERM PDE model can be used for computer-based experimentation, for example, parameter variation and changes in the model equations or alternate models, to enhance a quantitative understanding of a postulated tissue engineering and regenerative medicine procedure.

The author would welcome reader impressions of the computer-based TC models discussed in this book. Comments can be directed to wes1@lehigh.edu.

References

[1] C-S. Chou, et al., Spatial dynamics of multistage cell lineages in tissue stratification, Biophysical Journal 90 (November 2010) 3145–3154.
[2] https://www.nibib.nih.gov/science-education/science-topics/tissue-engineering-and-regenerative-medicine.
[3] K. Soetaert, J. Cash, F. Mazzia, Solving Differential Equations in R, Springer-Verlag, Heidelberg, Germany, 2012.

1

One PDE model development

Introduction

This book is directed to the computer-based modeling of *tissue engineering* and *regenerative medicine*. These areas offer the possibility of implantable, synthetic tissues and organs. If this research and development (R&D) works out (succeeds), the replacement of damaged and diseased tissues and organs will be feasible (and dependency on donor organs will end).

The computational methodology presented in the book provides quantitative analysis and understanding of this R&D in reaching the goal of producing synthetic tissues and organs. A principal feature of the book is the testing and documentation of a series of R routines that the reader/researcher/analyst can use to implement the PDE-based methodology.

The following additional background is from [4].

What are tissue engineering and regenerative medicine?

Tissue engineering evolved from the field of biomaterials development and refers to the practice of combining scaffolds, cells, and biologically active molecules into functional tissues. The goal of tissue engineering is to assemble functional constructs that restore, maintain, or improve damaged tissues or whole organs. Artificial skin and cartilage are examples of engineered tissues that have been approved by the FDA; however, currently they have limited use in human patients.

Regenerative medicine is a broad field that includes tissue engineering but also incorporates research on self-healing – where the body uses its own systems, sometimes with help foreign biological material to recreate cells and rebuild tissues and organs. The terms "tissue engineering" and "regenerative medicine" have become largely interchangeable, as the field hopes to focus on cures instead of treatments for complex, often chronic, diseases.

This field continues to evolve. In addition to medical applications, non-therapeutic applications include using tissues as biosensors to detect biological or chemical threat agents, and tissue chips that can be used to test the toxicity of an experimental medication.

The environment in which tissue engineering and regenerative medicine take place is explained in [4].

PDE Modeling of Tissue Engineering and Regenerative Medicine. https://doi.org/10.1016/B978-0-44-318740-7.00006-1

How do tissue engineering and regenerative medicine work?
Cells are the building blocks of tissue, and tissues are the basic unit of function in the body. Generally, groups of cells make and secrete their own support structures, called extra-cellular matrix. This matrix, or scaffold, does more than just support the cells; it also acts as a relay station for various signaling molecules. Thus, cells receive messages from many sources that become available from the local environment. Each signal can start a chain of responses that determine what happens to the cell. By understanding how individual cells respond to signals, interact with their environment, and organize into tissues and organisms, researchers have been able to manipulate these processes to mend damaged tissues or even create new ones.
The process often begins with building a scaffold from a wide set of possible sources, from proteins to plastics. Once scaffolds are created, cells with or without a "cocktail" of growth factors can be introduced. If the environment is right, a tissue develops. In some cases, the cells, scaffolds, and growth factors are all mixed together at once, allowing the tissue to "self-assemble."
Another method to create new tissue uses an existing scaffold. The cells of a donor organ are stripped and the remaining collagen scaffold is used to grow new tissue. This process has been used to bioengineer heart, liver, lung, and kidney tissue. This approach holds great promise for using scaffolding from human tissue discarded during surgery and combining it with a patient's own cells to make customized organs that would not be rejected by the immune system.

To begin the use of the PDE-based methodology, a model for O_2 transfer in the seeded stem cells is considered next [1].

1.1 O_2 transfer, 1D cylindrical coordinates

The tissue construct (TC=stem cells and scaffold) is modeled with cylindrical geometry. The cylindrical coordinates (r, θ, z) are initially reduced to 1D in r.

The transfer of O_2 to the stem cells in the TC is modeled by the following PDE ([2,3], eq. (1)) that is a material balance on O_2.[1]

$$\frac{\partial u_1}{\partial t} = D_{1,z} \frac{\partial^2 u_1}{\partial z^2} - q_1 \tag{1.1-1}$$

where

u_1	O_2 concentration in the TC
z	axial distance along the TC
t	time
q_1	volumetric rate of O_2 consumption in the TC
$D_{1,z}$	effective O_2 diffusivity

[1] The established convention is used here for naming PDE dependent variables as u_n where n is the number of the PDE.

Eq. (1.1-1) indicates $u_1 = u_1(z, t)$ (as derived in chapter Appendix A1).

Eq. (1.1-1) is first order in t and requires one initial condition (IC).

$$u_1(z, t = 0) = f_1(z) \tag{1.1-2}$$

$f_1(z)$ is a function to be specified.

Eq. (1.1-1) is second order in z and requires two boundary conditions (BCs).

$$u_1(z = 0, t) = g_1(t) \tag{1.1-3}$$

Eq. (1.1-3) is a Dirichlet BC.[2] $g_1(t)$ is a function that specifies the input O_2 concentration to the TC system.

$$\frac{\partial u_1(z = z_u, t)}{\partial z} = 0 \tag{1.1-4}$$

Eq. (1.1-4) is a Neumann BC[2]. z_u is the TC thickness in z and specifies a zero flux of O_2 at the cell-scaffold interface.

The O_2 source term, q_1, is given by Michaelis-Menten kinetics.

$$q_1 = p \frac{Q_m u_1}{c_m + u_1} \tag{1.1-5}$$

where p, Q_m, c_m are parameters to be specified.[3]

Eqs. (1.1) constitute the 1D PDE model for O_2 dynamics in the TC. The numerical solution to eqs. (1.1), $u_1(z, t)$, is computed in Chapter 2.

This model is next extended to 2D by adding a radial coordinate r in (r, θ, z).

1.2 O_2 transfer, 2D cylindrical coordinates

The O_2 balance in 2D cylindrical coordinates is (derived in chapter Appendix A1)

$$\frac{\partial u_1}{\partial t} = D_{1,r} \left(\frac{\partial^2 u_1}{\partial r^2} + \frac{1}{r} \frac{\partial u_1}{\partial r} \right) + D_{1,z} \frac{\partial^2 u_1}{\partial z^2} - q_1 \tag{1.2-1}$$

$D_{1,r}$, $D_{1,z}$ are effective diffusivities in r and z in the TC.

Eq. (1.2-1) is first order in t and requires one IC.

$$u_1(r, z, t = 0) = f_2(r, z) \tag{1.2-2}$$

$f_2(r, z)$ is a function to be specified.

---

[2] A Dirichlet BC specifies the dependent variable, $u_1(z, t)$, at the boundary. A Neumann BC specifies the first order spatial derivative of $u_1(z, t)$ with respect to z at the boundary. A Robin BC includes both the dependent variable and the first order spatial derivative, frequently as a linear combination, e.g., $D_{1,z} \dfrac{\partial u_1(z = z_u, t)}{\partial z} + k_m u_1(z = z_u, t) = 0$.

[3] p can be a function of t as determined from experimental data [3].

Eq. (1.2-1) is second order in r and z, and requires two BC in each of these spatial independent variables. For r, the BCs are homogeneous Neumann (no flux).

$$\frac{\partial u_1(r=0, z, t)}{\partial r} = 0 \tag{1.2-3}$$

$$\frac{\partial u_1(r=r_u, z, t)}{\partial r} = 0 \tag{1.2-4}$$

r_u is the outer boundary in r. BC (1.2-3) specifies symmetry at $r = 0$. BC (1.2-4) specifies zero flux at $r = r_u$.

For z, the BCs are Dirichlet at $z = z_l = 0$ and homogeneous Neumann at $z = z_u = 1$.

$$u_1(r, z = z_l = 0, t) = g_2(r, t) \tag{1.2-5}$$

$$\frac{\partial u_1(r, z = z_u, t)}{\partial z} = 0 \tag{1.2-6}$$

$g_2(r, t)$ is a function to be specified.

The numerical solution to eqs. (1.2), $u_1(r, z, t)$, is computed in Chapter 3.

1.3 Summary and conclusions

A one PDE model for the spatiotemporal distribution of O_2 in the tissue construct is formulated in 1D and 2D. The auxiliary conditions (ICs, BCs) complete the model. The implementation of this model as a series of source routines is considered in Chapters 2, 3.

Appendix A1 Derivation of O_2 balance

The derivation of eq. (1.1-1) is detailed first

A1.1 O_2 transfer, 1D cylindrical coordinates

Eq. (1.1-1) is based on a material (mass) balance for an incremental volume $A_c \Delta z$.

$$A_c \Delta z \frac{\partial u_1}{\partial t} = -A_c D_{1,z} \frac{\partial u_1}{\partial z}|_z - (-A_c D_{1,z} \frac{\partial u_1}{\partial z}|_{z+\Delta z}) - A_c \Delta z q_1 \tag{A.1-1}$$

The diffusive flux into and out of the incremental volume is based on Fick's first law (also Fourier's first law).

$$q_z = -D_{1,z} \frac{\partial u_1}{\partial z} \tag{A.1-2}$$

with $q_z > 0$ for $\frac{\partial u_1}{\partial z} < 0$ (diffusion in the direction of decreasing $u_1(z, t)$).

Division of eq. (A.1-1) by $A_c \Delta z$ and minor rearrangement gives

$$\frac{\partial u_1}{\partial t} = D_{1,z} \left(\frac{\frac{\partial u_1}{\partial z}|_{z+\Delta z} - \frac{\partial u_1}{\partial z}|_z}{\Delta z} \right) - q_1 \tag{A.1-3}$$

With $\Delta z \to 0$, eq. (A.1-3) is eq. (1.1-1).

Eqs. (A.1) constitute a 1D model for the TC in a cylindrical geometry. Extension to 2D is considered next.

A1.2 O_2 transfer, 2D cylindrical coordinates

The following material (mass) balance for an incremental volume $2\pi r \Delta r \Delta z$ is analogous to eq. (A.1-1).

$$2\pi r \Delta r \Delta z \frac{\partial u_1}{\partial t} =$$

$$-2\pi r \Delta z D_{1,r} \frac{\partial u_1}{\partial r}|_r - \left(-2\pi r \Delta z D_{1,r} \frac{\partial u_1}{\partial r}|_{r+\Delta r} \right)$$

$$-2\pi r \Delta r D_{1,z} \frac{\partial u_1}{\partial z}|_z - \left(-2\pi r \Delta r D_{1,z} \frac{\partial u_1}{\partial z}|_{z+\Delta z} \right)$$

$$-2\pi r \Delta r \Delta z q_1 \tag{A.2-1}$$

Division of eq. (A.2-1) by $2\pi r \Delta r \Delta z$ and minor rearrangement gives

$$\frac{\partial u_1}{\partial t} =$$

$$D_{1,r} \frac{1}{r} \left(\frac{r \frac{\partial u_1}{\partial r}|_{r+\Delta r} - r \frac{\partial u_1}{\partial r}|_r}{\Delta r} \right)$$

$$+D_{1,z} \left(\frac{\frac{\partial u_1}{\partial z}|_{z+\Delta z} - \frac{\partial u_1}{\partial z}|_z}{\Delta z} \right)$$

$$-q_1 \tag{A.2-2}$$

With $\Delta r \to 0$, $\Delta z \to 0$,

$$\frac{\partial u_1}{\partial t} =$$

$$D_{1,r} \frac{1}{r} \frac{\partial \left(r \frac{\partial u_1}{\partial r} \right)}{\partial r}$$

$$+D_{1,z} \frac{\partial^2 u_1}{\partial z^2}$$

$$-q_1 \tag{A.2-3}$$

With expansion of the radial group,

$$\frac{\partial u_1}{\partial t} =$$

$$D_{1,r}\left(\frac{\partial^2 u_1}{\partial r^2} + \frac{1}{r}\frac{\partial u_1}{\partial r}\right)$$

$$+D_{1,z}\frac{\partial^2 u_1}{\partial z^2}$$

$$-q_1 \qquad\qquad\qquad\qquad (A.2\text{-}4)$$

Eq. (A.2-4) is eq. (1.2-1).

References

[1] B.D. MacArthur, P.S. Stumpf, R.O.C. Oreffo, From mathematical modeling and machine learning to clinical reality, in: Principles of Tissue Engineering, fifth edition, Elsevier, Cambridge, MA, 2020, Chapter 2.
[2] J. Malda, J. Rouwkema, D.E. Martens, E.P. le Comte, F.K. Kooy, J. Tramper, C.A. van Blitterswij, J. Riesle, Oxygen gradients in tissue-engineered PEGT/PBT cartilaginous constructs: measurement and modeling, Biotechnology and Bioengineering 86 (1) (2004) 9–18.
[3] B. Obradovic, J.H. Meldon, L.E. Freed, G. Vunjak-Novakovic, Glycosaminoglycan deposition in engineered cartilage: experiments and mathematical model, AIChE Journal 46 (9) (2000) 1860–1871.
[4] https://www.nibib.nih.gov/science-education/science-topics/tissue-engineering-and-regenerative-medicine.

2

1D PDE model implementation

Introduction

The 1D PDE model of eqs. (1.1) in Chapter 1 is implemented in this chapter as a set of routines programmed in R.[1]

2.1 1D PDE model implementation

Eqs. (1.1) that constitute the 1D PDE model for the O_2 concentration in the stem cells seeded on a scaffold are implemented with the following R routines, starting with a main program.

2.1.1 Main program, test cases

The main program for eqs. (1.1) follows.

```
#
# 1D TC model
#
# Delete previous workspaces
  rm(list=ls(all=TRUE))
#
# Access ODE integrator
  library("deSolve");
#
# Access functions for numerical solution
  setwd("f:/tissue engineering/chap2");
  source("pde1a.R");
#
# Parameters
  D1z=0.1;
  p=1;
  Qm=1;
  cm=1;
  pQm=p*Qm;
```

[1] R is a quality, open source, scientific computing system available from the Internet [1].

7

```
  u10=1;
  ncase=1;
#
# Spatial grid (in z)
  nz=21;zl=0;zu=1;dz=(zu-zl)/(nz-1);dz2=dz^2;
  z=seq(from=zl,to=zu,by=dz);
#
# Independent variable for ODE integration
  t0=0;tf=1;nout=11;
  tout=seq(from=t0,to=tf,by=(tf-t0)/(nout-1));
#
# Initial condition (t=0)
  u0=rep(0,nz);
  if(ncase==1){
    for(iz in 1:nz){
      u0[iz]=u10;
    }
  }
  if(ncase==2){
    for(iz in 1:nz){
      u0[iz]=sin(pi*(iz-1)/(nz-1));
    }
  }
  ncall=0;
#
# ODE integration
  out=lsodes(y=u0,times=tout,func=pde1a,
      sparsetype ="sparseint",rtol=1e-6,
      atol=1e-6,maxord=5);
  nrow(out)
  ncol(out)
#
# Array for plotting numerical solution
  u1=matrix(0,nrow=nz,ncol=nout);
  for(it in 1:nout){
    for(iz in 1:nz){
      u1[iz,it]=out[it,iz+1];
    }
  }
#
# Display numerical solution
  iv=seq(from=1,to=nout,by=2);
```

```
  for(it in iv){
    cat(sprintf("\n    t      z      u1(z,t)\n"));
    iv=seq(from=1,to=nz,by=2);
    for(iz in iv){
      cat(sprintf("%6.1f%6.1f%12.3e\n",
          tout[it],z[iz],u1[iz,it]));
    }
  }
#
# Calls to ODE routine
  cat(sprintf("\n\n ncall = %5d\n\n",ncall));
#
# Plot PDE solution
#
# u1(z,t)
# 2D
  par(mfrow=c(1,1));
  matplot(x=z,y=u1,type="l",xlab="z",ylab="u1(z,t)",
    xlim=c(zl,zu),lty=1,main="",lwd=2,col="black");
#
# 3D
  persp(z,tout,u1,theta=45,phi=30,
        xlim=c(zl,zu),ylim=c(t0,tf),xlab="z",
        ylab="t",zlab="u1(z,t)");
```

Listing 2.1: Main program for eqs. (1.1).

We can note the following details about Listing 2.1.

- Previous workspaces are deleted.

```
#
# 1D TC model
#
# Delete previous workspaces
  rm(list=ls(all=TRUE))
```

- The R ODE integrator library deSolve is accessed [1].

```
#
# Access ODE integrator
  library("deSolve");
#
# Access functions for numerical solution
```

```
setwd("f:/tissue engineering/chap2");
source("pde1a.R");
```

Then the directory with the files for the solution of eqs. (1.1) is designated. Note that setwd (set working directory) uses / rather than the usual \. The ordinary differential equation/method of lines (ODE/MOL)[2] routine, pde1a.R, is accessed through setwd.
• The model parameters are specified numerically.

```
#
# Parameters
  D1z=0.1;
  p=1;
  Qm=1;
  cm=1;
  pQm=p*Qm;
  u10=1;
  ncase=1;
```

where
  • D1z: effective O_2 diffusivity.
  • p,Qm,cm: parameters in q of eq. (1.1-5).
  • u10: initial condition (IC) for eq. (1.1-1).
  • ncase: index for selection of the initial condition function of eq. (1.1-2) (explained subsequently).
• A spatial grid for eq. (1.1-1) is defined with 21 points so that z = 0,1/20=0.05,...,1. The stem cell length is a normalized value, $z = z_u = 1$.

```
#
# Spatial grid (in z)
  nz=21;zl=0;zu=1;dz=(zu-zl)/(nz-1);dz2=dz^2;
  z=seq(from=zl,to=zu,by=dz);
```

• An interval in t is defined for 11 output points, so that tout=0,1/10=0.1,...,1. The time scale is normalized with $t_f = 1$ specified as the final time that is considered appropriate, e.g., minutes, days.

```
#
# Independent variable for ODE integration
  t0=0;tf=1;nout=11;
  tout=seq(from=t0,to=tf,by=(tf-t0)/(nout-1));
```

---

[2]The method of lines is a general numerical algorithm for PDEs in which the boundary value (spatial) derivatives are replaced with algebraic approximations, in this case, finite differences (FDs). The resulting system of initial value ODEs is then integrated (solved) with a library ODE integrator.

- IC (1.1-2) is implemented for `ncase=1,2`.

```
#
# Initial condition (t=0)
  u0=rep(0,nz);
  if(ncase==1){
    for(iz in 1:nz){
      u0[iz]=u10;
    }
  }
  if(ncase==2){
    for(iz in 1:nz){
      u0[iz]=sin(pi*(iz-1)/(nz-1));
    }
  }
  ncall=0;
```

For `ncase=1`, the IC function is a constant, `u10`. For `ncase=2`, the IC function is a sine function (discussed subsequently). Also, the counter for the calls to `pde1a` is initialized.

- The system of nz=21 ODEs is integrated by the library integrator `lsodes` (available in deSolve, [1]). As expected, the inputs to `lsodes` are the ODE function, `pde1a`, the IC vector `u0`, and the vector of output values of t, `tout`. The length of `u0` (21) informs `lsodes` how many ODEs are to be integrated. `func,y,times` are reserved names.

```
#
# ODE integration
  out=lsodes(y=u0,times=tout,func=pde1a,
      sparsetype ="sparseint",rtol=1e-6,
      atol=1e-6,maxord=5);
  nrow(out)
  ncol(out)
```

`nrow,ncol` confirm the dimensions of `out`.

- $u_1(z,t)$ is placed in a matrix for subsequent plotting.

```
#
# Array for plotting numerical solution
  u1=matrix(0,nrow=nz,ncol=nout);
  for(it in 1:nout){
    for(iz in 1:nz){
      u1[iz,it]=out[it,iz+1];
    }
  }
```

The offset +1 is required because the first element of the solution vectors in out is the value of t and the 2 to 22 elements are the 21 values of u_1. These dimensions from the preceding calls to nrow,ncol are confirmed in the subsequent output.

- The numerical values of $u_1(z,t)$ returned by lsodes are displayed. Every second value in t and every second value in z appear from by=2,2.

```
#
# Display numerical solution
  iv=seq(from=1,to=nout,by=2);
  for(it in iv){
    cat(sprintf("\n    t     z      u1(z,t)\n"));
    iv=seq(from=1,to=nz,by=2);
    for(iz in iv){
      cat(sprintf("%6.1f%6.1f%12.3e\n",
          tout[it],z[iz],u1[iz,it]));
    }
  }
```

- The number of calls to pde1a is displayed at the end of the solution.

```
#
# Calls to ODE routine
  cat(sprintf("\n\n ncall = %5d\n\n",ncall));
```

- $u_1(z,t)$ is plotted in 2D against z and parametrically in t with the R utility matplot. $u_1(z,t)$ is plotted in 3D with R utility persp. par(mfrow=c(1,1)) specifies a 1×1 matrix of plots, that is, one plot on a page.

```
#
# Plot PDE solution
#
# u1(z,t)
# 2D
  par(mfrow=c(1,1));
  matplot(x=z,y=u1,type="l",xlab="z",ylab="u1(z,t)",
    xlim=c(zl,zu),lty=1,main="",lwd=2,col="black");
#
# 3D
  persp(z,tout,u1,theta=45,phi=30,
        xlim=c(zl,zu),ylim=c(t0,tf),xlab="z",
        ylab="t",zlab="u1(z,t)");
```

This completes the discussion of the main program for eqs. (1.1). The ODE/MOL routine pde1a called by lsodes from the main program for the numerical MOL integration of eqs. (1.1) is next.

2.1.2 ODE/MOL routine

pde1a called in the main program of Listing 2.1 follows.

```
  pde1a=function(t,u,parm){
#
# Function pde1a computes the t derivatives
# of u1(z,t)
#
# One vector to one vector
  u1=rep(0,nz);
  for(iz in 1:nz){
    u1[iz]=u[iz];
  }
#
# u1zz, Neumann BC, z=0,zu
  u1zz=rep(0,nz);
  for(iz in 1:nz){
    if(iz==1){
      u1zz[1]=2*(u1[2]-u1[1])/dz2;}
    if(iz==nz){
      u1zz[nz]=2*(u1[nz-1]-u1[nz])/dz2;}
    if((iz>1)&(iz<nz)){
      u1zz[iz]=(u1[iz+1]-2*u1[iz]+u1[iz-1])/dz2;}
  }
#
# u1t
  u1t=rep(0,nz);
  for(iz in 1:nz){
#
#   Michaelis-Menten
    q=pQm*u1[iz]/(cm+u1[iz]);
#
#   PDE
    u1t[iz]=D1z*u1zz[iz]-q;
  }
#
# 1D to 1D vector
  ut=rep(0,nz);
  for(iz in 1:nz){
    ut[iz]=u1t[iz];
  }
```

```
#
# Increment calls to pde1a
  ncall <<- ncall+1;
#
# Return derivative vector
  return(list(c(ut)));
  }
```

Listing 2.2: ODE/MOL routine for eqs. (1.1).

We can note the following details about Listing 2.2.

- The function is defined.

```
  pde1a=function(t,u,parm){
#
# Function pde1a computes the t derivatives
# of u1(z,t)
```

t is the current value of t in eqs. (1.1). u is the 21-vector of ODE/PDE dependent variables. parm is an argument to pass parameters to pde1a (unused, but required in the argument list). The arguments must be listed in the order stated to properly interface with lsodes called in the main program of Listing 2.1. The derivative vector of the LHS of eq. (1.1-1) is calculated and returned to lsodes as explained subsequently.
- The input vector u is placed in a vector u1 to facilitate the programming of eq. (1.1-1).

```
#
# One vector to one vector
  u1=rep(0,nz);
  for(iz in 1:nz){
    u1[iz]=u[iz];
  }
```

- The derivative $\dfrac{\partial^2 u_1}{\partial z^2}$ in eq. (1.1-1) is programmed.

```
#
# u1zz, Neumann BC, z=0,zu
  u1zz=rep(0,nz);
  for(iz in 1:nz){
    if(iz==1){
      u1zz[1]=2*(u1[2]-u1[1])/dz2;}
    if(iz==nz){
      u1zz[nz]=2*(u1[nz-1]-u1[nz])/dz2;}
    if((iz>1)&(iz<nz)){
```

```
    u1zz[iz]=(u1[iz+1]-2*u1[iz]+u1[iz-1])/dz2;}
}
```

This coding requires some additional explanation.

- The derivative $\dfrac{\partial^2 u_1}{\partial z^2}$ in eq. (1.1-1) is approximated with a three point centered finite difference (FD).

$$\frac{\partial^2 u_1(z,t)}{\partial z^2} \approx \frac{(u_1(z+\Delta z,t) - 2u_1(z,t) + u_1(z-\Delta z,t))}{\Delta z^2} + O(\Delta z^2) \qquad (2.1\text{-}1)$$

$O(\Delta z^2)$ indicates that the error in the FD approximation is second order in Δz. The variation in the numerical solution of eq. (1.1-1) can be studied as a function of the FD increment Δz by varying nz in Listing 2.1.

- At $z = z_l = 0$, a homogeneous Neumann BC is used in place of BC (1.1-3) for a special case test to validate the coding in pde1a as explained subsequently.

$$\frac{\partial u_1(z=z_l,t)}{\partial z} \approx \frac{\partial u_1(z_l + \Delta z,t) - u_1(z_l - \Delta z,t)}{2\Delta z} = 0$$

so that the fictitious value at $z = z_l - \Delta z$ is

$$u_1(z_l - \Delta z,t) = u_1(z_l + \Delta z,t)$$

Eq. (2.1-1) at $z = z_l$ is therefore

$$\frac{\partial^2 u_1(z=z_l,t)}{\partial z^2} \approx 2\frac{(u_1(z_l + \Delta z,t) - u_1(z_l,t))}{\Delta z^2} + O(\Delta z^2) \qquad (2.1\text{-}2)$$

and is programmed as

```
#
# u1zz, Neumann BC, z=0,zu
  u1zz=rep(0,nz);
  for(iz in 1:nz){
    if(iz==1){
      u1zz[1]=2*(u1[2]-u1[1])/dz2;}
```

- Similarly, at $z = z_u$, BC (1.1-4) is used to eliminate the fictitious value $u_1(z_u + \Delta z,t)$, and eq. (2.1-1) is

$$\frac{\partial^2 u_1(z_u,t)}{\partial z^2} \approx \frac{2(u_1(z_u - \Delta z,t) - u_1(z_u,t))}{\Delta z^2} \qquad (2.1\text{-}3)$$

and is programmed as

```
    if(iz==nz){
      u1zz[nz]=2*(u1[nz-1]-u1[nz])/dz2;}
```

- For the interior points $z_l < z < z_u$, eq. (2.1-1) is programmed as

```
if((iz>1)&(iz<nz)){
   u1zz[iz]=(u1[iz+1]-2*u1[iz]+u1[iz-1])/dz2;}
}
```

  The final } concludes the for in z.

  $\dfrac{\partial^2 u_1}{\partial z^2}$ is now available for use in the programming of eq. (1.1-1).

- Eq. (1.1-1) is programmed in the MOL format.

```
#
# u1t
  u1t=rep(0,nz);
  for(iz in 1:nz){
#
#   Michaelis-Menten
    q=pQm*u1[iz]/(cm+u1[iz]);
#
#   PDE
    u1t[iz]=D1z*u1zz[iz]-q;
  }
```

- The 21 ODE derivatives are placed in the vector ut for return to lsodes to take the next step in t along the solution.

```
#
# One vector to one vector
  ut=rep(0,nz);
  for(iz in 1:nz){
    ut[iz]=u1t[iz];}
```

- The counter for the calls to pde1a is incremented and returned to the main program of Listing 2.1 by <<-.

```
#
# Increment calls to pde1a
  ncall <<- ncall+1;
```

- The vector ut is returned as a list as required by lsodes. c is the R vector utility.

```
#
# Return derivative vector
  return(list(c(ut)));
  }
```

The final } concludes pde1a.

This completes the discussion of pde1a. The output from the main program of Listing 2.1 and ODE/MOL routine pde1a of Listing 2.2 is considered next.

2.1.3 Numerical, graphical output

For ncase=1 in the main program of Listing 2.1, the output is

Table   2.1 Numerical output from Listings 2.1, 2.2, ncase=1.

[1] 11

[1] 22

t	z	u1(z,t)
0.0	0.0	1.000e+00
0.0	0.1	1.000e+00
0.0	0.2	1.000e+00
0.0	0.3	1.000e+00
0.0	0.4	1.000e+00
0.0	0.5	1.000e+00
0.0	0.6	1.000e+00
0.0	0.7	1.000e+00
0.0	0.8	1.000e+00
0.0	0.9	1.000e+00
0.0	1.0	1.000e+00

t	z	u1(z,t)
0.2	0.0	9.025e-01
0.2	0.1	9.025e-01
0.2	0.2	9.025e-01
0.2	0.3	9.025e-01
0.2	0.4	9.025e-01
0.2	0.5	9.025e-01
0.2	0.6	9.025e-01
0.2	0.7	9.025e-01
0.2	0.8	9.025e-01
0.2	0.9	9.025e-01
0.2	1.0	9.025e-01

.   .

.   .

.   .

Output for t = 0.4,0.6, 0.8 removed

continued on next page

Table 2.1   (continued)

	.	.
	.	.
	.	.

t	z	u1(z,t)
1.0	0.0	5.671e-01
1.0	0.1	5.671e-01
1.0	0.2	5.671e-01
1.0	0.3	5.671e-01
1.0	0.4	5.671e-01
1.0	0.5	5.671e-01
1.0	0.6	5.671e-01
1.0	0.7	5.671e-01
1.0	0.8	5.671e-01
1.0	0.9	5.671e-01
1.0	1.0	5.671e-01

ncall =   46

We can note the following details about this output.

- 11 t output points as the first dimension of the solution matrix out from lsodes as programmed in the main program of Listing 2.1 (with nout=11).
- The solution matrix out returned by lsodes has 22 elements as a second dimension. The first element is the value of t. Elements 2 to 22 are $u_1(z,t)$ (for each of the 21 output points).
- The solution is displayed for t=0,1/10=0.1,...,1 as programmed in Listing 2.1 (every second value of t is displayed as explained previously).
- The solution is displayed for z=0,1/20=0.05,...,1 as programmed in Listing 2.1 (every second value of z is displayed as explained previously).
- IC (1.1-2) is confirmed ($t = 0$).
- The solution $u_1(z,t)$ does not vary with z since (1) the IC (1.1-2), $u_1(z,t = 0) = 1$, and (2) homogeneous Neumann (no flux) BCs at $z = z_l$, $z = z_u$ are consistent with the IC. This is an important check since a variation of the solution with z would indicate a programming error.
- The decay in the solution with t results from the source term $q > 0$ in eq. (1.1-1). $q = 0$ in Listing 2.1 would result in a constant solution that does not change from the IC $u_1(z,t = 0) = 1$. This is also an important special case that is left as an exercise.
- The computational effort as indicated by ncall = 46 is modest so that lsodes computed the solution to eqs. (1.1) efficiently.

The graphical output is in Figs. 2.1.
The solution in Fig. 2.1-1 does not vary in z and decays in t (from $q > 0$).

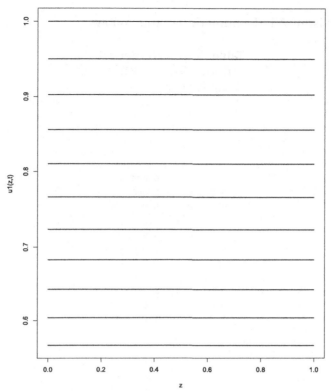

FIGURE 2.1-1 $u_1(z, t)$ from eqs. (1.1), 2D, ncase=1.

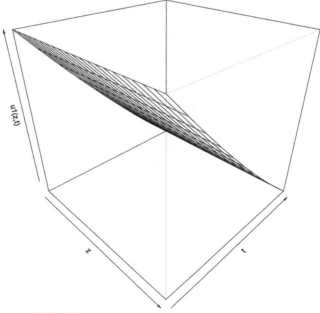

FIGURE 2.1-2 $u_1(z, t)$ from eqs. (1.1), 3D, ncase=1.

For ncase=2 in Listing 2.1, the numerical output follows.

Table 2.2 Numerical output from Listings 2.1, 2.2, ncase=2.

[1] 11

[1] 22

t	z	u1(z,t)
0.0	0.0	0.000e+00
0.0	0.1	3.090e-01
0.0	0.2	5.878e-01
0.0	0.3	8.090e-01
0.0	0.4	9.511e-01
0.0	0.5	1.000e+00
0.0	0.6	9.511e-01
0.0	0.7	8.090e-01
0.0	0.8	5.878e-01
0.0	0.9	3.090e-01
0.0	1.0	1.225e-16

t	z	u1(z,t)
0.2	0.0	3.795e-01
0.2	0.1	4.162e-01
0.2	0.2	5.099e-01
0.2	0.3	6.209e-01
0.2	0.4	7.071e-01
0.2	0.5	7.392e-01
0.2	0.6	7.071e-01
0.2	0.7	6.209e-01
0.2	0.8	5.099e-01
0.2	0.9	4.162e-01
0.2	1.0	3.795e-01

.   .
.   .
.   .

Output for t = 0.4,0.6, 0.8 removed

.   .
.   .
.   .

t	z	u1(z,t)
1.0	0.0	3.168e-01
1.0	0.1	3.178e-01
1.0	0.2	3.205e-01

continued on next page

Table 2.2   (continued)

1.0	0.3	3.237e-01
1.0	0.4	3.264e-01
1.0	0.5	3.274e-01
1.0	0.6	3.264e-01
1.0	0.7	3.237e-01
1.0	0.8	3.205e-01
1.0	0.9	3.178e-01
1.0	1.0	3.168e-01

```
ncall =    133
```

We can note the following details about this output.

- IC (1.1-2) is confirmed (at $t = 0$).

t	z	u1(z,t)
0.0	0.0	0.000e+00
0.0	0.1	3.090e-01
0.0	0.2	5.878e-01
0.0	0.3	8.090e-01
0.0	0.4	9.511e-01
0.0	0.5	1.000e+00
0.0	0.6	9.511e-01
0.0	0.7	8.090e-01
0.0	0.8	5.878e-01
0.0	0.9	3.090e-01
0.0	1.0	1.225e-16

Note in particular the value of $u_1(z = 0.5, t)$.

- The solution remains symmetric around $z = 0.5$. This is an important test since any asymmetry would indicate a programming error.
- The solution decays with increasing t for $q > 0$ in eq. (1.1-1).
- The homogeneous Neumann BCs are verified and the solution approaches a steady state value of approximately 0.32.
- The computational effort as indicated by `ncall` = 133 is modest so that `lsodes` computed the solution to eqs. (1.1) efficiently.

The graphical output is in Figs. 2.2.

This solution displays the properties discussed for Table 2.2 (symmetry around $z = 0.5$ and approach to a steady state solution).

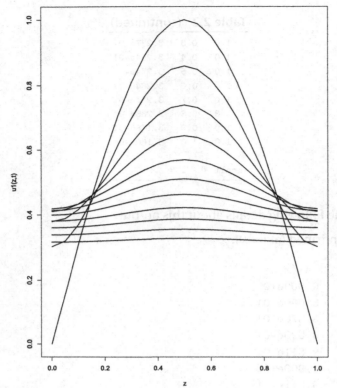

FIGURE 2.2-1 $u_1(z,t)$ from eqs. (1.1), 2D, ncase=2.

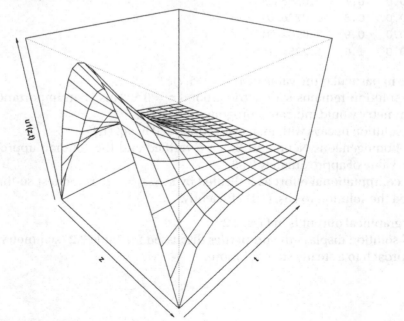

FIGURE 2.2-2 $u_1(z,t)$ from eqs. (1.1), 3D, ncase=2.

This concludes the discussion of Listings 2.1, 2.2 for which homogeneous Neumann BCs are imposed at $z = z_l, z_u$ (to test various solution properties with `ncase=1,2` in the coding as discussed previously). The case with a Dirichlet BC at $z = z_l$, as specified by $g_1(t)$ in eq. (1.1-3), is now considered.

2.1.4 Main program, application

The main program with a Dirichlet BC at $z = z_l$ follows.

```
#
# 1D TC model
#
# Delete previous workspaces
  rm(list=ls(all=TRUE))
#
# Access ODE integrator
  library("deSolve");
#
# Access functions for numerical solution
  setwd("f:/tissue engineering/chap2");
  source("pde1b.R");
#
# Parameters
  D1z=0.1;
  p=0.1;
  Qm=1;
  cm=1;
  pQm=p*Qm;
  u1b=1;
  u10=0;
#
# Spatial grid (in z)
  nz=21;zl=0;zu=1;dz=(zu-zl)/(nz-1);dz2=dz^2;
  z=seq(from=zl,to=zu,by=dz);
#
# Independent variable for ODE integration
  t0=0;tf=1;nout=11;
  tout=seq(from=t0,to=tf,by=(tf-t0)/(nout-1));
#
# Initial condition (t=0)
  u0=rep(0,nz);
```

```
  for(iz in 1:nz){
    u0[iz]=u10;
  }
  ncall=0;
#
# ODE integration
  out=lsodes(y=u0,times=tout,func=pde1b,
      sparsetype ="sparseint",rtol=1e-6,
      atol=1e-6,maxord=5);
  nrow(out)
  ncol(out)
#
# Array for plotting numerical solution
  u1=matrix(0,nrow=nz,ncol=nout);
  for(it in 1:nout){
    for(iz in 1:nz){
      u1[iz,it]=out[it,iz+1];
     }
  u1[1,it]=u1b;
  }
#
# Display numerical solution
  iv=seq(from=1,to=nout,by=2);
  for(it in iv){
    cat(sprintf("\n    t      z      u1(z,t)\n"));
    iv=seq(from=1,to=nz,by=2);
    for(iz in iv){
      cat(sprintf("%6.1f%6.1f%12.3e\n",
          tout[it],z[iz],u1[iz,it]));
    }
  }
#
# Calls to ODE routine
  cat(sprintf("\n\n ncall = %5d\n\n",ncall));
#
# Plot PDE solutions
#
# u1(z,t)
# 2D
  par(mfrow=c(1,1));
  matplot(x=z[2:nz],y=u1[2:nz,],type="l",xlab="z",ylab="u1(z,t)",
    xlim=c(zl,zu),lty=1,main="",lwd=2,col="black");
```

```
#
# 3D
  persp(z,tout,u1,theta=45,phi=30,
        xlim=c(zl,zu),ylim=c(t0,tf),xlab="z",
        ylab="t",zlab="u1(z,t)");
```

Listing 2.3: Main program for eqs. (1.1), Dirichlet BC.

We can note the following details about Listing 2.3 (with some repetition of the discussion of Listing 2.1 so that the explanation is self contained).

• Previous workspaces are deleted.

```
#
# 1D TC model
#
# Delete previous workspaces
  rm(list=ls(all=TRUE))
```

• The R ODE integrator library deSolve is accessed [1].

```
#
# Access ODE integrator
  library("deSolve");
#
# Access functions for numerical solution
  setwd("f:/tissue engineering/chap2");
  source("pde1b.R");
```

Then the directory with the files for the solution of eqs. (1.1) is designated. Note that setwd (set working directory) uses / rather than the usual \. The ordinary differential equation/method of lines (ODE/MOL) routine, pde1b, is accessed through setwd.

• The model parameters are specified numerically.

```
#
# Parameters
  D1z=0.1;
  p=0.1;
  Qm=1;
  cm=1;
  pQm=p*Qm;
  u1b=1;
  u10=0;
```

In particular,

- u1b: Boundary value $u_1(z = 0, t) = u_{1b} = 1$ in BC (1.1-3).
- u10: Initial value $u_1(z, t = 0) = u_{10} = 0$ in IC (1.1-2).
- A spatial grid for eq. (1.1-1) is defined with 21 points so that z = 0,1/20=0.05,...,1. The stem cell length is a normalized value, $z = z_u = 1$.

```
#
# Spatial grid (in z)
  nz=21;zl=0;zu=1;dz=(zu-zl)/(nz-1);dz2=dz^2;
  z=seq(from=zl,to=zu,by=dz);
```

- An interval in t is defined for 11 output points, so that tout=0,1/10=0.1,...,1. The time scale is normalized with $t_f = 1$ specified as the final time that is considered appropriate, e.g., minutes, days.

```
#
# Independent variable for ODE integration
  t0=0;tf=1;nout=11;
  tout=seq(from=t0,to=tf,by=(tf-t0)/(nout-1));
```

- IC (1.1-2) is implemented.

```
#
# Initial condition (t=0)
  u0=rep(0,nz);
  for(iz in 1:nz){
    u0[iz]=u10;
  }
  ncall=0;
```

Also, the counter for the calls to pde1b is initialized.
- The system of nz=21 ODEs is integrated by the library integrator lsodes (available in deSolve, [1]). As expected, the inputs to lsodes are the ODE function, pde1b, the IC vector u0, and the vector of output values of t, tout. The length of u0 (21) informs lsodes how many ODEs are to be integrated. func,y,times are reserved names.

```
#
# ODE integration
  out=lsodes(y=u0,times=tout,func=pde1b,
      sparsetype ="sparseint",rtol=1e-6,
      atol=1e-6,maxord=5);
  nrow(out)
  ncol(out)
```

nrow,ncol confirm the dimensions of out.

- $u_1(z, t)$ is placed in a matrix for subsequent plotting.

```
#
# Array for plotting numerical solution
  u1=matrix(0,nrow=nz,ncol=nout);
  for(it in 1:nout){
    for(iz in 1:nz){
      u1[iz,it]=out[it,iz+1];
      }
  u1[1,it]=u1b;
  }
```

The offset +1 is required since the first element of the solution vectors in out is the value of *t* and the 2 to 22 elements are the 21 values of u_1. These dimensions from the preceding calls to nrow,ncol are confirmed in the subsequent output. The BC $u_1(z = 0, t) = u_{1b} = 1$ is algebraic, but pde1b returns only solutions to ODEs. Therefore, this boundary value is restated for the numerical output.

- The numerical values of $u_1(z, t)$ returned by lsodes are displayed. Every second value in *t* and every second value in *z* appear from by=2,2.

```
#
# Display numerical solution
  iv=seq(from=1,to=nout,by=2);
  for(it in iv){
    cat(sprintf("\n    t     z      u1(z,t)\n"));
    iv=seq(from=1,to=nz,by=2);
    for(iz in iv){
      cat(sprintf("%6.1f%6.1f%12.3e\n",
          tout[it],z[iz],u1[iz,it]));
    }
  }
```

- The number of calls to pde1b is displayed at the end of the solution.

```
#
# Calls to ODE routine
  cat(sprintf("\n\n ncall = %5d\n\n",ncall));
```

- $u_1(z, t)$ is plotted in 2D against *z* and parametrically in *t* with the R utility matplot, and in 3D with the R utility persp. par(mfrow=c(1,1)) specifies a 1×1 matrix of plots, that is, one plot on a page.

```
#
# Calls to ODE routine
```

```
    cat(sprintf("\n\n ncall = %5d\n\n",ncall));
#
# Plot PDE solutions
#
# u1(z,t)
# 2D
  par(mfrow=c(1,1));
  matplot(x=z[2:nz],y=u1[2:nz,],type="l",xlab="z",ylab="u1(z,t)",
    xlim=c(zl,zu),lty=1,main="",lwd=2,col="black");
#
# 3D
    persp(z,tout,u1,theta=45,phi=30,
        xlim=c(zl,zu),ylim=c(t0,tf),xlab="z",
        ylab="t",zlab="u1(z,t)");
```

This completes the discussion of the main program for eqs. (1.1). The ODE/MOL routine pde1b called by lsodes from the main program of Listing 2.3 for the numerical MOL integration of eqs. (1.1) is next.

2.1.5 ODE/MOL routine

```
  pde1b=function(t,u,parm){
#
# Function pde1b computes the t derivatives
# of u1(z,t)
#
# One vector to one vector
  u1=rep(0,nz);
  for(iz in 1:nz){
    u1[iz]=u[iz];
  }
# u1zz, Dirichlet BC at z=zl, Neumann BC, z=zu
  u1zz=rep(0,nz);
  for(iz in 1:nz){
    if(iz==1){
      u1[1]=u1b;}
    if(iz==nz){
      u1zz[nz]=2*(u1[nz-1]-u1[nz])/dz2;}
    if((iz>1)&(iz<nz)){
      u1zz[iz]=(u1[iz+1]-2*u1[iz]+u1[iz-1])/dz2;}
  }
```

```
#
# u1t
  u1t=rep(0,nz);
  for(iz in 2:nz){
#
#   Michaelis-Menten
    q=pQm*u1[iz]/(cm+u1[iz]);
#
#   PDE
    u1t[iz]=D1z*u1zz[iz]-q;
  }
  u1t[1]=0;
#
# 1D to 1D vector
  ut=rep(0,nz);
  for(iz in 1:nz){
    ut[iz]=u1t[iz];
  }
#
# Increment calls to pde1b
  ncall <<- ncall+1;
#
# Return derivative vector
  return(list(c(ut)));
  }
```

Listing 2.4: ODE/MOL routine for eqs. (1.1), Dirichlet BC.

We can note the following details about Listing 2.2 (with some repetition of the discussion of Listing 2.2 so that the explanation is self contained).

- The function is defined.

```
    pde1b=function(t,u,parm){
#
# Function pde1b computes the t derivatives
# of u1(z,t)
```

t is the current value of *t* in eqs. (1.1). u is the 21-vector of ODE/PDE dependent variables. parm is an argument to pass parameters to pde1b (unused, but required in the argument list). The arguments must be listed in the order stated to properly interface with lsodes called in the main program of Listing 2.3. The derivative vector of the LHS of eq. (1.1-1) is calculated and returned to lsodes as explained subsequently.

- The input vector u is placed in a vector u1 to facilitate the programming of eq. (1.1-1).

```
#
# One vector to one vector
  u1=rep(0,nz);
  for(iz in 1:nz){
    u1[iz]=u[iz];
  }
```

- The derivative $\frac{\partial^2 u_1}{\partial z^2}$ in eq. (1.1-1) is programmed.

```
#
# u1zz, Dirichlet BC at z=zl, Neumann BC, z=zu
  u1zz=rep(0,nz);
  for(iz in 1:nz){
    if(iz==1){
      u1[1]=u1b;}
    if(iz==nz){
      u1zz[nz]=2*(u1[nz-1]-u1[nz])/dz2;}
    if((iz>1)&(iz<nz)){
      u1zz[iz]=(u1[iz+1]-2*u1[iz]+u1[iz-1])/dz2;}
  }
```

This coding requires some additional explanation.

- The derivative $\frac{\partial^2 u_1}{\partial z^2}$ in eq. (1.1-1) is approximated with a three point centered finite difference (FD).

$$\frac{\partial^2 u_1(z,t)}{\partial z^2} \approx \frac{(u_1(z+\Delta z,t) - 2u_1(z,t) + u_1(z-\Delta z,t))}{\Delta z^2} + O(\Delta z^2) \qquad (2.2\text{-}1)$$

$O(\Delta z^2)$ indicates that the error in the FD approximation is second order in Δz. The variation in the numerical solution of eq. (1.1-1) can be studied as a function of the FD increment Δz by varying nz in Listing 2.3.

- Dirichlet BC (1.1-3) is applied at $z = z_l = 0$.

$$g_1(t) = u_{1b} = 1 \qquad (2.2\text{-}2)$$

and is programmed as

```
for(iz in 1:nz){
  if(iz==1){
    u1[1]=u1b;}
```

- At $z = z_u$, Neumann BC (1.1-4) is used to eliminate the fictitious value $u_1(z_u + \Delta z, t)$, and eq. (2.1-1) is

$$\frac{\partial^2 u_1(z_u, t)}{\partial z^2} \approx \frac{2(u_1(z_u - \Delta z, t) - u_1(z_u, t))}{\Delta z^2} \qquad (2.2\text{-}3)$$

  that is programmed as

```
if(iz==nz){
  u1zz[nz]=2*(u1[nz-1]-u1[nz])/dz2;}
```

- For the interior points $z_l < z < z_u$, eq. (2.1-1) is programmed as

```
if((iz>1)&(iz<nz)){
  u1zz[iz]=(u1[iz+1]-2*u1[iz]+u1[iz-1])/dz2;}
}
```

  The final } concludes the for in z.
  $\frac{\partial^2 u_1}{\partial z^2}$ is now available for use in the programming of eq. (1.1-1).
- Eq. (1.1-1) is programmed in the MOL format.

```
#
# u1t
  u1t=rep(0,nz);
  for(iz in 2:nz){
#
#    Michaelis-Menten
    q=pQm*u1[iz]/(cm+u1[iz]);
#
#    PDE
    u1t[iz]=D1z*u1zz[iz]-q;
  }
  u1t[1]=0;
```

$\frac{\partial u_1(z = z_l, t)}{\partial t} = 0$ so that $u_1(z = z_l, t)$ remains at the boundary value (eq. (1.1-3)).
- The 21 ODE derivatives are placed in the vector ut for return to lsodes to take the next step in t along the solution.

```
#
# 1D to 1D vector
  ut=rep(0,nz);
  for(iz in 1:nz){
    ut[iz]=u1t[iz];
  }
```

- The counter for the calls to pde1b is incremented and returned to the main program of Listing 2.3 by <<-.

```
#
# Increment calls to pde1b
  ncall <<- ncall+1;
```

- The vector ut is returned as a list as required by lsodes. c is the R vector utility.

```
#
# Return derivative vector
  return(list(c(ut)));
  }
```

The final } concludes pde1b.

This completes the discussion of pde1b. The output from the main program of Listing 2.3 and ODE/MOL routine pde1b of Listing 2.4 is considered next.

2.1.6 Numerical, graphical output

Table 2.3 Numerical output from Listings 2.3, 2.4, Dirichlet BC.

[1] 11

[1] 22

t	z	u1(z,t)
0.0	0.0	1.000e+00
0.0	0.1	0.000e+00
0.0	0.2	0.000e+00
0.0	0.3	0.000e+00
0.0	0.4	0.000e+00
0.0	0.5	0.000e+00
0.0	0.6	0.000e+00
0.0	0.7	0.000e+00
0.0	0.8	0.000e+00
0.0	0.9	0.000e+00
0.0	1.0	0.000e+00

t	z	u1(z,t)
0.2	0.0	1.000e+00
0.2	0.1	6.138e-01

continued on next page

Table 2.3 (continued)

t	z	u1(z,t)
0.2	0.2	3.149e-01
0.2	0.3	1.335e-01
0.2	0.4	4.659e-02
0.2	0.5	1.340e-02
0.2	0.6	3.195e-03
0.2	0.7	6.358e-04
0.2	0.8	1.065e-04
0.2	0.9	1.532e-05
0.2	1.0	3.680e-06
.	.	
.	.	
.	.	

Output for t = 0.4,0.6,
0.8 removed

t	z	u1(z,t)
.	.	
.	.	
.	.	
t	z	u1(z,t)
1.0	0.0	1.000e+00
1.0	0.1	8.149e-01
1.0	0.2	6.424e-01
1.0	0.3	4.889e-01
1.0	0.4	3.587e-01
1.0	0.5	2.536e-01
1.0	0.6	1.732e-01
1.0	0.7	1.151e-01
1.0	0.8	7.669e-02
1.0	0.9	5.497e-02
1.0	1.0	4.797e-02

ncall =   142

We can note the following details about this output.

- IC (1.1-2) is confirmed (at $t = 0$).
  Note in particular the value of $u_1(z = 0, t = 0)$.
- BC (1.1-3) is verified ($u_1(z = 0, t) = g_1(t) = 1$).

The graphical output is in Figs. 2.3.

This solution displays the properties discussed for Table 2.3 (BC (1.1-3) that moves the solution from IC (1.1-2)).

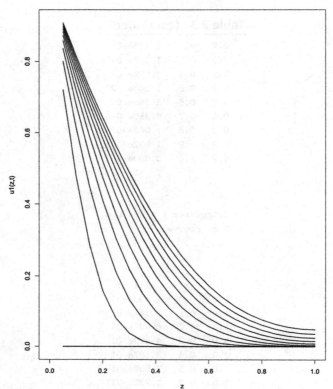

FIGURE 2.3-1 $u_1(z, t)$ from eqs. (1.1), 2D, Dirichlet BC.

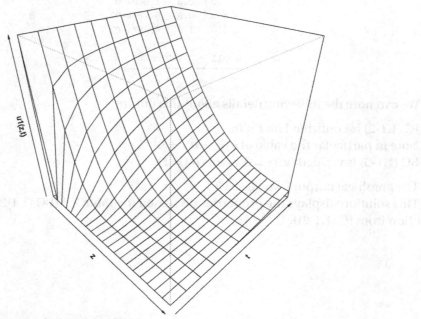

FIGURE 2.3-2 $u_1(z, t)$ from eqs. (1.1), 3D, Dirichlet BC.

This concludes the discussion of Listings 2.3, 2.4 for which a Dirichlet BC is specified at $z = z_l$ and a Neumann BC at $z = z_u$.

2.2 Summary and conclusions

A 1D O_2 diffusion model is implemented in this chapter for: (1) two Neumann BCs, (2) one Dirichlet and one Neumann BC applied to the spatial domain of the stem cells seeded on a scaffold. The R routines are based on the MOL with the spatial derivatives approximated by finite differences. For (2), the increase of O_2 concentration provides an essential condition for the stem cell differentiation that could be the basis for tissue engineering and regenerative medicine.

The 1D model of this chapter is extended to a 2D PDE model in cylindrical coordinates in Chapter 3.

Reference

[1] K. Soetaert, J. Cash, F. Mazzia, Solving Differential Equations in R, Springer-Verlag, Heidelberg, Germany, 2012.

3

$$\blacksquare\blacksquare\blacksquare$$

2D PDE model implementation

Introduction

The 2D PDE model of eqs. (1.2) in Chapter 1 is implemented in this chapter as a set of routines programmed in R,[1] starting with a main program.

3.1 2D PDE model implementation

Eqs. (1.2) that constitute the 2D PDE model for the O_2 concentration in the stem cells seeded on a scaffold are implemented with the following R routines, starting with a main program.

3.1.1 Main program, test cases

```
#
# 2D TC model
#
# Delete previous workspaces
  rm(list=ls(all=TRUE))
#
# Access ODE integrator
  library("deSolve");
#
# Access functions for numerical solution
  setwd("f:/tissue engineering/chap3");
  source("pde2a.R");
#
# Parameters
  D1r=0.1;
  D1z=0.1;
  p=1;
  Qm=1;
  cm=1;
  pQm=p*Qm;
```

---

[1] R is a quality, open source, scientific computing system available from the Internet [1].

PDE Modeling of Tissue Engineering and Regenerative Medicine. https://doi.org/10.1016/B978-0-44-318740-7.00008-5

```
  u10=1;
  ncase=1;
#
# Spatial grid (in r)
  nr=11;rl=0;ru=1;dr=(ru-rl)/(nr-1);dr2=dr^2;
  r=seq(from=rl,to=ru,by=dr);
#
# Spatial grid (in z)
  nz=21;zl=0;zu=1;dz=(zu-zl)/(nz-1);dz2=dz^2;
  z=seq(from=zl,to=zu,by=dz);
#
# Independent variable for ODE integration
  t0=0;tf=1;nout=11;
  tout=seq(from=t0,to=tf,by=(tf-t0)/(nout-1));
#
# Initial condition (t=0)
  u0=rep(0,nr*nz);
  if(ncase==1){
    for(iz in 1:nz){
    for(ir in 1:nr){
      u0[(iz-1)*nr+ir]=u10;
    }
    }
  }
  if(ncase==2){
    for(iz in 1:nz){
    for(ir in 1:nr){
      u0[(iz-1)*nr+ir]=
        sin(pi*(ir-1)/(nr-1))*
        sin(pi*(iz-1)/(nz-1));
    }
    }
  }
  ncall=0;
#
# ODE integration
  out=lsodes(y=u0,times=tout,func=pde2a,
      sparsetype ="sparseint",rtol=1e-6,
      atol=1e-6,maxord=5);
  nrow(out)
  ncol(out)
```

```
#
# Array for plotting numerical solution
  u1=matrix(0,nrow=nz,ncol=nout);
  for(it in 1:nout){
    cat(sprintf("\n      t        z      (iz-1)nr+ir  u1(r=0,z,t)\n"));
    ir=1;
    for(iz in 1:nz){
      u1[iz,it]=out[it,(iz-1)*nr+ir+1];
      cat(sprintf("%7.1f%9.3f%9.1f%16.3e\n",
          tout[it],z[iz],(iz-1)*nr+ir,u1[iz,it]));
      }
  }
#
# Calls to ODE routine
  cat(sprintf("\n\n ncall = %5d\n\n",ncall));
#
# Plot PDE solutions
#
# u1(z,t)
# 2D
  par(mfrow=c(1,1));
  matplot(x=z,y=u1,type="l",xlab="z",ylab="u1(r=0,z,t)",
    lty=1,main="",lwd=2,col="black");
#
# 3D
  persp(z,tout,u1,theta=45,phi=30,xlim=c(zl,zu),ylim=c(t0,tf),
    xlab="z",ylab="t",zlab="u1(r=0,z,t)");
```

Listing 3.1: Main program for eqs. (1.2).

We can note the following details about Listing 3.1 (with some repetition of the discussion of Listing 2.1 so that the following explanation is self contained).

- Previous workspaces are deleted.

```
#
# 2D TC model
#
# Delete previous workspaces
  rm(list=ls(all=TRUE))
```

- The R ODE integrator library deSolve is accessed [1].

```
#
# Access ODE integrator
  library("deSolve");
#
# Access functions for numerical solution
  setwd("f:/tissue engineering/chap3");
  source("pde2a.R");
```

Then the directory with the files for the solution of eqs. (1.2) is designated. Note that setwd (set working directory) uses / rather than the usual \. The ordinary differential equation/method of lines (ODE/MOL)[2] routine, pde2a.R, is accessed through setwd.
- The model parameters are specified numerically.

```
#
# Parameters
  D1r=0.1;
  D1z=0.1;
  p=1;
  Qm=1;
  cm=1;
  pQm=p*Qm;
  u10=1;
  ncase=1;
```

where
- D1r, D1z: effective O_2 diffusivities.
- p,Qm,cm: parameters in q of eq. (1.1-5).
- u10: initial condition (IC) for eq. (1.2-1).
- ncase: index for selection of the initial condition function of eq. (1.2-2) (explained subsequently).
- A spatial grid in r for eq. (1.2-1) is defined with 11 points so that $r = 0, 1/10 = 0.1, \ldots, 1$. The stem cell length is a normalized value, $r = r_u = 1$.

```
#
# Spatial grid (in r)
  nr=11;rl=0;ru=1;dr=(ru-rl)/(nr-1);dr2=dr^2;
  r=seq(from=rl,to=ru,by=dr);
```

---

[2] The method of lines is a general numerical algorithm for PDEs in which the boundary value (spatial) derivatives are replaced with algebraic approximations, in this case, finite differences (FDs). The resulting system of initial value ODEs is then integrated (solved) with a library ODE integrator.

- A spatial grid in z for eq. (1.2-1) is defined with 21 points so that z = 0,1/20=0.05,...,1. The stem cell length is a normalized value, $z = z_u = 1$.

```
#
# Spatial grid (in z)
  nz=21;zl=0;zu=1;dz=(zu-zl)/(nz-1);dz2=dz^2;
  z=seq(from=zl,to=zu,by=dz);
```

- An interval in t is defined for 11 output points, so that tout=0,1/10=0.1,...,1. The time scale is normalized with $t_f = 1$ specified as the final time that is considered appropriate, e.g., minutes, days.

```
#
# Independent variable for ODE integration
  t0=0;tf=1;nout=11;
  tout=seq(from=t0,to=tf,by=(tf-t0)/(nout-1));
```

- IC (1.2-2) is implemented for ncase=1,2.

```
#
# Initial condition (t=0)
  u0=rep(0,nr*nz);
  if(ncase==1){
    for(iz in 1:nz){
    for(ir in 1:nr){
      u0[(iz-1)*nr+ir]=u10;
    }
    }
  }
  if(ncase==2){
    for(iz in 1:nz){
    for(ir in 1:nr){
      u0[(iz-1)*nr+ir]=
        sin(pi*(ir-1)/(nr-1))*
        sin(pi*(iz-1)/(nz-1));
    }
    }
  }
  ncall=0;
```

For ncase=1, the IC function is a constant, u10. For ncase=2, the IC function is a sine function (discussed subsequently). Also, the counter for the calls to pde2a is initialized.
- The system of nz=11*21=231 ODEs is integrated by the library integrator lsodes (available in deSolve, [1]). As expected, the inputs to lsodes are the ODE function, pde2a, the

IC vector u0, and the vector of output values of t, tout. The length of u0 (231) informs lsodes how many ODEs are to be integrated. func, y, times are reserved names.

```
#
# ODE integration
  out=lsodes(y=u0,times=tout,func=pde2a,
      sparsetype ="sparseint",rtol=1e-6,
      atol=1e-6,maxord=5);
  nrow(out)
  ncol(out)
```

nrow, ncol confirm the dimensions of out.

- $u_1(r, z, t)$ is placed in a matrix for subsequent plotting.

```
#
# Array for plotting numerical solution
  u1=matrix(0,nrow=nz,ncol=nout);
  for(it in 1:nout){
    cat(sprintf("\n       t       z       (iz-1)nr+ir   u1(r=0,z,t)\n"));
    ir=1;
    for(iz in 1:nz){
      u1[iz,it]=out[it,(iz-1)*nr+ir+1];
      cat(sprintf("%7.1f%9.3f%9.1f%16.3e\n",
          tout[it],z[iz],(iz-1)*nr+ir,u1[iz,it]));
    }
  }
```

Since $u_1(r, z, t)$ is a function of three variables, r, z, t, it cannot be plotted in 3D. Therefore, the solution at $r = 0$ is selected with ir=1, that is, $u_1(r = 0, z, t)$ is plotted.

The offset +1 is required because the first element of the solution vectors in out is the value of t and the 2 to 232 elements are the 231 values of u_1. These dimensions from the preceding calls to nrow, ncol are confirmed in the subsequent output.

- The number of calls to pde2a is displayed at the end of the solution.

```
#
# Calls to ODE routine
  cat(sprintf("\n\n ncall = %5d\n\n",ncall));
```

- par(mfrow=c(1,1)) specifies a 1×1 matrix of plots, that is, one plot on a page. $u_1(r = 0, z, t)$ is plotted in 2D against z and parametrically in t with the R utility matplot, and in 3D against z and t with the R utility persp.

```
#
# Plot PDE solutions
#
# u1(r,z,t)
# 2D
  par(mfrow=c(1,1));
  matplot(x=z,y=u1,type="l",xlab="z",ylab="u1(r=0,z,t)",
    lty=1,main="",lwd=2,col="black");
#
# 3D
  persp(z,tout,u1,theta=45,phi=30,xlim=c(zl,zu),ylim=c(t0,tf),
    xlab="z",ylab="t",zlab="u1(r=0,z,t)");
```

This completes the discussion of the main program for eqs. (1.2). The ODE/MOL routine pde2a called by lsodes from the main program for the numerical MOL integration of eqs. (1.2) is next.

3.1.2 ODE/MOL routine

pde2a called in the main program of Listing 3.1 follows.

```
  pde2a=function(t,u,parm){
#
# Function pde2a computes the t derivative
# of u1(r,z,t)
#
# 1D to 2D vector
  u1=matrix(0,nrow=nr,ncol=nz);
  for(iz in 1:nz){
  for(ir in 1:nr){
    u1[ir,iz]=u[(iz-1)*nr+ir];
  }
  }
#
# u1rr, Neumann BC, r=0,ru
  u1rr=matrix(0,nrow=nr,ncol=nz);
  for(iz in 1:nz){
  for(ir in 1:nr){
    if(ir==1){
      u1rr[1,iz]=4*(u1[2,iz]-u1[1,iz])/dr2;}
    if(ir==nr){
      u1rr[nr,iz]=2*(u1[nr-1,iz]-u1[nr,iz])/dr2;}
```

```
    if((ir>1)&(ir<nr)){
      u1rr[ir,iz]=(u1[ir+1,iz]-2*u1[ir,iz]+u1[ir-1,iz])/dr2+
                  (1/r[ir])*(u1[ir+1,iz]-u1[ir-1,iz])/(2*dr);}
  }
  }
#
# u1zz, Neumann BC, z=0,zu
  u1zz=matrix(0,nrow=nr,ncol=nz);
  for(iz in 1:nz){
  for(ir in 1:nr){
    if(iz==1){
      u1zz[ir,1]=2*(u1[ir,2]-u1[ir,1])/dz2;}
    if(iz==nz){
      u1zz[ir,nz]=2*(u1[ir,nz-1]-u1[ir,nz])/dz2;}
    if((iz>1)&(iz<nz)){
      u1zz[ir,iz]=(u1[ir,iz+1]-2*u1[ir,iz]+u1[ir,iz-1])/dz2;}
  }
  }
#
# u1t
  u1t=matrix(0,nrow=nr,ncol=nz);
  for(iz in 1:nz){
  for(ir in 1:nr){
#
#   Michaelis-Menten
    q=pQm*u1[ir,iz]/(cm+u1[ir,iz]);
#
#   PDE
    u1t[ir,iz]=D1r*u1rr[ir,iz]+D1z*u1zz[ir,iz]-q;
  }
  }
#
# 2D to 1D vector
  ut=rep(0,nr*nz);
  for(iz in 1:nz){
  for(ir in 1:nr){
    ut[(iz-1)*nr+ir]=u1t[ir,iz];
  }
  }
#
```

```
# Increment calls to pde2a
  ncall <<- ncall+1;
#
# Return derivative vector
  return(list(c(ut)));
  }
```

Listing 3.2: ODE/MOL routine for eqs. (1.2).

We can note the following details about Listing 3.2 (with some repetition of the discussion of Listing 2.2 so that the following explanation is self contained).

• The function is defined.

```
  pde2a=function(t,u,parm){
#
# Function pde2a computes the t derivative
# of u1(r,z,t)
```

t is the current value of t in eqs. (1.2). u is the 231-vector of ODE/PDE dependent variables. parm is an argument to pass parameters to pde2a (unused, but required in the argument list). The arguments must be listed in the order stated to properly interface with lsodes called in the main program of Listing 3.1. The derivative vector of the LHS of eq. (1.2-1) is calculated and returned to lsodes as explained subsequently.

• The input vector u is placed in a matrix u1 to facilitate the programming of eq. (1.2-1).

```
#
# 1D vector to 2D matrix
  u1=matrix(0,nrow=nr,ncol=nz);
  for(iz in 1:nz){
  for(ir in 1:nr){
    u1[ir,iz]=u[(iz-1)*nr+ir];
  }
  }
```

• The radial derivative group $\dfrac{\partial^2 u_1}{\partial r^2} + \dfrac{1}{r}\dfrac{\partial u_1}{\partial r}$ in eq. (1.2-1) is programmed.

```
#
# u1rr, Neumann BC, r=0,ru
  u1rr=matrix(0,nrow=nr,ncol=nz);
  for(iz in 1:nz){
  for(ir in 1:nr){
    if(ir==1){
      u1rr[1,iz]=4*(u1[2,iz]-u1[1,iz])/dr2;}
```

```
    if(ir==nr){
      u1rr[nr,iz]=2*(u1[nr-1,iz]-u1[nr,iz])/dr2;}
    if((ir>1)&(ir<nr)){
      u1rr[ir,iz]=(u1[ir+1,iz]-2*u1[ir,iz]+u1[ir-1,iz])/dr2+
                  (1/r[ir])*(u1[ir+1,iz]-u1[ir-1,iz])/(2*dr);}
  }
  }
```

This coding requires some additional explanation.

- The radial derivative term $\dfrac{1}{r}\dfrac{\partial u_1}{\partial r}$ in eq. (1.2-1) is indeterminate at $r = 0$ with the application of BC (1.2-3).

$$\frac{1}{r}\frac{\partial u_1}{\partial r}\Big|_{r=0} = \frac{0}{0}$$

- Application of l'Hospital's rule to this indeterminate term gives

$$\frac{1}{r}\frac{\partial u_1}{\partial r}\Big|_{r=0} = \frac{\partial^2 u_1}{\partial r^2}\Big|_{r=0}$$

and the radial group in eq. (1.2-1) is

$$\frac{\partial^2 u_1}{\partial r^2} + \frac{1}{r}\frac{\partial u_1}{\partial r}\Big|_{r=0}$$
$$= 2\frac{\partial^2 u_1}{\partial r^2}$$

- The FD approximation of the second derivative at $r = 0$ is

$$\frac{\partial^2 u_1}{\partial r^2}\Big|_{r=0}$$
$$\approx \frac{u_1(r = \Delta r, z, t) - 2u(r = 0, z, t) + u_1(r = -\Delta r, z, t)}{\Delta r^2}$$

- The fictitious term $u_1(r = -\Delta r, z, t)$ is approximated from BC (1.2-3) with a two point FD centered at $r = 0$

$$\frac{u_1(r = \Delta r, z, t) - u_1(r = -\Delta r, z, t)}{2\Delta r} = 0$$

or

$$u_1(r = -\Delta r, z, t) \approx u_1(r = \Delta r, z, t)$$

- The second derivative is then approximated as

$$\frac{\partial^2 u_1}{\partial r^2}\Big|_{r=0}$$
$$\approx 2\frac{u_1(r = \Delta r, z, t) - u(r = 0, z, t)}{\Delta r^2}$$

- The radial group in eq. (1.2-1) at $r = 0$ is therefore

$$\frac{\partial^2 u_1}{\partial r^2} + \frac{1}{r}\frac{\partial u_1}{\partial r}\Big|_{r=0}$$

$$\approx 4\frac{u_1(r = \Delta r, z, t) - u_1(r = 0, z, t)}{\Delta r^2}$$

which is programmed as

```
#
# u1rr, Neumann BC, r=0,ru
  u1rr=matrix(0,nrow=nr,ncol=nz);
  for(iz in 1:nz){
  for(ir in 1:nr){
    if(ir==1){
      u1rr[1,iz]=4*(u1[2,iz]-u1[1,iz])/dr2;}
```

- The radial derivative group $\dfrac{\partial^2 u_1}{\partial r^2} + \dfrac{1}{r}\dfrac{\partial u_1}{\partial r}$ in eq. (1.2-1) at $r = r_u = 1$ with application of homogeneous Neumann BC (1.2-4) is

$$\frac{\partial^2 u_1}{\partial r^2} + \frac{1}{r}\frac{\partial u_1}{\partial r}\Big|_{r=r_u=1} = \frac{\partial^2 u_1}{\partial r^2}$$

As before (at $r = 0$), this derivative is approximated as

$$\frac{\partial^2 u_1}{\partial r^2}\Big|_{r=r_u=1} \approx 2\frac{u_1(r = r_u - \Delta r, z, t) - u(r = r_u, z, t)}{\Delta r^2}$$

and is programmed as

```
    if(ir==nr){
      u1rr[nr,iz]=2*(u1[nr-1,iz]-u1[nr,iz])/dr2;}
```

which includes BC (1.2-4).
- For the interior points $r > r_l$, $r < r_u$, the radial group in eq. (1.2-1) is approximated with FDs as

$$\frac{\partial^2 u_1}{\partial r^2} + \frac{1}{r}\frac{\partial u_1}{\partial r}$$

$$\approx \frac{u_1(r + \Delta r, z, t) - 2u_1(r, z, t) + u_1(r - \Delta r, z, t)}{\Delta r^2}$$

$$+\frac{1}{r}\frac{u_1(r + \Delta r, z, t) - u_1(r - \Delta r, z, t)}{2\Delta r}$$

and is programmed as

```
        if((ir>1)&(ir<nr)){
        u1rr[ir,iz]=(u1[ir+1,iz]-2*u1[ir,iz]+u1[ir-1,iz])/dr2+
                    (1/r[ir])*(u1[ir+1,iz]-u1[ir-1,iz])/(2*dr);}
  }
  }
```

This completes the programming of the radial group in eq. (1.2-1) over the 2D $n_r \times n_z$ grid in r, z. This group can now be used in the MOL programming of eq. (1.2-1) as indicated subsequently.

- The axial derivative $\frac{\partial^2 u_1}{\partial z^2}$ in eq. (1.2-1) is programmed.

```
#
# u1zz, Neumann BC, z=0,zu
  u1zz=matrix(0,nrow=nr,ncol=nz);
  for(iz in 1:nz){
  for(ir in 1:nr){
    if(iz==1){
    u1zz[ir,1]=2*(u1[ir,2]-u1[ir,1])/dz2;}
    if(iz==nz){
    u1zz[ir,nz]=2*(u1[ir,nz-1]-u1[ir,nz])/dz2;}
    if((iz>1)&(iz<nz)){
    u1zz[ir,iz]=(u1[ir,iz+1]-2*u1[ir,iz]+u1[ir,iz-1])/dz2;}
  }
  }
```

This programming follows analogously from the previous coding in r with homogeneous Neumann BCs at $z = z_l$, $z = z_u$

```
z=zl

    if(iz==1){
    u1zz[ir,1]=2*(u1[ir,2]-u1[ir,1])/dz2;}

z=zu
    if(iz==nz){
    u1zz[ir,nz]=2*(u1[ir,nz-1]-u1[ir,nz])/dz2;}
```

- The MOL programming of eq. (1.2-1) gives the derivative $\frac{\partial u_1(r,z,t)}{\partial t}$, including the Michaelis-Menten source term q.

```
#
# u1t
  u1t=matrix(0,nrow=nr,ncol=nz);
```

```
    for(iz in 1:nz){
    for(ir in 1:nr){
#
#   Michaelis-Menten
    q=pQm*u1[ir,iz]/(cm+u1[ir,iz]);
#
#   PDE
    u1t[ir,iz]=D1r*u1rr[ir,iz]+D1z*u1zz[ir,iz]-q;
    }
    }
```

- The 231 ODE derivatives are placed in the vector ut for return to lsodes to take the next step in *t* along the solution.

```
#
# 2D matrix to 1D vector
  ut=rep(0,nr*nz);
  for(iz in 1:nz){
  for(ir in 1:nr){
    ut[(iz-1)*nr+ir]=u1t[ir,iz];
  }
  }
```

- The counter for the calls to pde2a is incremented and returned to the main program of Listing 3.1 by <<-.

```
#
# Increment calls to pde2a
  ncall <<- ncall+1;
```

- The vector ut is returned as a list as required by lsodes. c is the R vector utility.

```
#
# Return derivative vector
  return(list(c(ut)));
  }
```

The final } concludes pde2a.

This completes the discussion of pde2a. The output from the main program of Listing 3.1 and ODE/MOL routine pde2a of Listing 3.2 is considered next.

3.1.3 Numerical, graphical output

Abbreviated output for ncase=1 (set in Listing 3.1) follows.

Table 3.1 Numerical output from List-ings 3.1, 3.2, ncase=1.

[1] 11

[1] 232

t	z	(iz-1)nr+ir	u1(r=0,z,t)
0.0	0.000	1.0	1.000e+00
0.0	0.050	12.0	1.000e+00
0.0	0.100	23.0	1.000e+00

.
.
.

Output for (iz-1)nr+ir = 34,...
188 removed

.
.
.

0.0	0.900	199.0	1.000e+00
0.0	0.950	210.0	1.000e+00
0.0	1.000	221.0	1.000e+00

t	z	(iz-1)nr+ir	u1(r=0,z,t)
0.1	0.000	1.0	9.506e-01
0.1	0.050	12.0	9.506e-01
0.1	0.100	23.0	9.506e-01

.
.
.

Output for (iz-1)nr+ir = 34,...
188 removed

.
.
.

0.1	0.900	199.0	9.506e-01
0.1	0.950	210.0	9.506e-01
0.1	1.000	221.0	9.506e-01

.
.
.

Output for t = 0.2,...0.9 removed

.
.
.

continued on next page

Table 3.1   (continued)

t	z	(iz-1)nr+ir	u1(r=0,z,t)
1.0	0.000	1.0	5.671e-01
1.0	0.050	12.0	5.671e-01
1.0	0.100	23.0	5.671e-01
.	.		.
.	.		.
.	.		.

```
     Output for (iz-1)nr+ir = 34,...
              188 removed
```

.	.		.
.	.		.
.	.		.
1.0	0.900	199.0	5.671e-01
1.0	0.950	210.0	5.671e-01
1.0	1.000	221.0	5.671e-01

```
ncall =   260
```

We can note the following details about this output.

- 11 t output points as the first dimension of the solution matrix out from lsodes as programmed in the main program of Listing 3.1 (with nout=11).
- The solution matrix out returned by lsodes has 232 elements as a second dimension. The first element is the value of t. Elements 2 to 232 are $u_1(r, z, t)$ from eqs. (1.2) (for the spatial grid with $n_r = 11, n_z = 21, (n_r)(n_z) = (11)(21) = 231$ points).
- The solution is displayed for t=0,1/10=0.1,...,1 as programmed in Listing 3.1.
- The solution $u_1(r = 0, z, t)$ is displayed for the index (iz-1)*nr+ir=1,12,...,221 with iz=1,...,21, ir=1 programmed in Listing 3.1.
- IC (1.2-2) is confirmed ($t = 0$).
- The solution $u_1(r = 0, z, t)$ does not vary with z since the IC $u_1(r, z, t = 0) = 1$ and BCs (1.2-3,4,5,6) are consistent (with BC (1.2-5) replaced with a homogeneous Neumann BC as programmed in pde2a in Listing 3.2). This is an important check since a variation of the solution with z would indicate a programming error.
- The decay in the solution with t results from the source term $q > 0$ in eq. (1.2-1). $q = 0$ in Listings 3.1, 3.2 would result in a constant solution that does not change from the IC $u_1(r = 0, z, t = 0) = 1$. This is also an important special case that is left as an exercise.
- The computational effort as indicated by ncall = 260 is modest so that lsodes computed the solution to eqs. (1.2) efficiently.

The graphical output is in Figs. 3.1.
The solution in Fig. 3.1-1 does not vary in z and decays from $q > 0$.

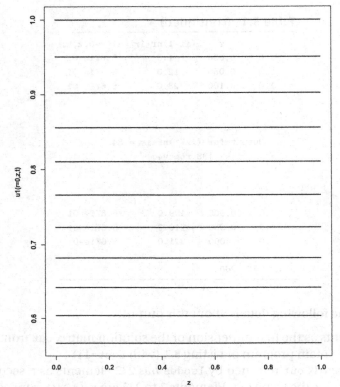

FIGURE 3.1-1 $u_1(r = 0, z, t)$ from eqs. (1.2), 2D, ncase=1.

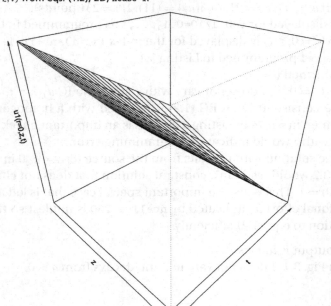

FIGURE 3.1-2 $u_1(r = 0, z, t)$ from eqs. (1.2), 3D, ncase=1.

For ncase=2 the numerical output follows.

Table 3.2   Numerical output from Listings 3.1, 3.2, ncase=2.

```
[1] 11

[1] 232

        t       z      (iz-1)nr+ir   u1(r=0,z,t)
       0.0    0.000        1.0         0.000e+00
       0.0    0.050       12.0         0.000e+00
       0.0    0.100       23.0         0.000e+00
                .                          .
                .                          .
                .                          .
         Output for (iz-1)nr+ir = 34,...
                   188 removed
                .                          .
                .                          .
                .                          .
       0.0    0.900      199.0         0.000e+00
       0.0    0.950      210.0         0.000e+00
       0.0    1.000      221.0         0.000e+00

        t       z      (iz-1)nr+ir   u1(r=0,z,t)
       0.1    0.000        1.0         1.450e-01
       0.1    0.050       12.0         1.533e-01
       0.1    0.100       23.0         1.768e-01
                .                          .
                .                          .
                .                          .
         Output for (iz-1)nr+ir = 34,...
                   188 removed
                .                          .
                .                          .
                .                          .
       0.1    0.900      199.0         1.768e-01
       0.1    0.950      210.0         1.533e-01
       0.1    1.000      221.0         1.450e-01
                .                          .
                .                          .
                .                          .
        Output for t = 0.2,...0.9 removed
                .                          .
                .                          .
                .                          .
```

continued on next page

Table 3.2 (continued)

t	z	(iz-1)nr+ir	u1(r=0,z,t)
1.0	0.000	1.0	2.138e-01
1.0	0.050	12.0	2.140e-01
1.0	0.100	23.0	2.145e-01
.	.		. .
.	.		.
.	.		.

```
       Output for (iz-1)nr+ir = 34,...
                188 removed
```

.	.		.
.	.		.
.	.		.
1.0	0.900	199.0	2.145e-01
1.0	0.950	210.0	2.140e-01
1.0	1.000	221.0	2.138e-01

```
ncall =   345
```

We can note the following details about this output.

- IC (1.2-2) is confirmed, $u_1(r = 0, z, t = 0) = 0$.
- The solution remains symmetric around $z = 0.5$. This is an important test since any asymmetry would indicate a programming error.
- The homogeneous Neumann BCs are verified and the solution approaches a steady state value of approximately 0.21.
- The computational effort as indicated by `ncall = 345` is modest so that `lsodes` computed the solution to eqs. (1.2) efficiently.

The graphical output is in Figs. 3.2.

This solution displays the properties discussed for Table 3.2. (symmetry around $z = 0.5$ and approach to a steady state solution).

This concludes the discussion of Listings 3.1, 3.2 for which homogeneous Neumann BCs are imposed at $r = r_l, r_u, z = z_l, z_u$ (to test various solution properties with `ncase=1,2` in the coding as discussed previously). The case with a Dirichlet BC at $z = z_l$, as specified by $g_2(r, t)$ in eq. (1.2-5), is now considered.

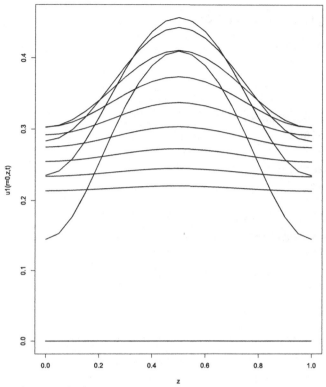

FIGURE 3.2-1 $u_1(r = 0, z, t)$ from eqs. (1.2), 2D, ncase=2.

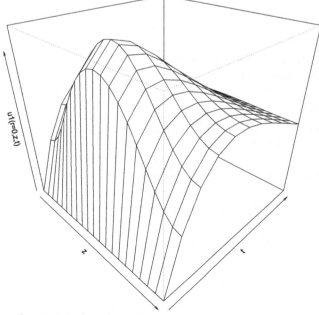

FIGURE 3.2-2 $u_1(r = 0, z, t)$ from eqs. (1.2), 3D, ncase=2.

3.1.4 Main program, application

The main program follows.

```
#
# 2D TC model
#
# Delete previous workspaces
  rm(list=ls(all=TRUE))
#
# Access ODE integrator
  library("deSolve");
#
# Access functions for numerical solution
  setwd("f:/tissue engineering/chap3");
  source("pde2b.R");
#
# Parameters
  D1r=0.1;
  D1z=0.1;
  p=1;
  Qm=1;
  cm=1;
  pQm=p*Qm;
  u1b=1;
  u10=0;
#
# Spatial grid (in r)
  nr=11;rl=0;ru=1;dr=(ru-rl)/(nr-1);dr2=dr^2;
  r=seq(from=rl,to=ru,by=dr);
#
# Spatial grid (in z)
  nz=21;zl=0;zu=1;dz=(zu-zl)/(nz-1);dz2=dz^2;
  z=seq(from=zl,to=zu,by=dz);
#
# Independent variable for ODE integration
  t0=0;tf=1;nout=11;
  tout=seq(from=t0,to=tf,by=(tf-t0)/(nout-1));
#
# Initial condition (t=0)
  u0=rep(0,nr*nz);
  for(iz in 1:nz){
  for(ir in 1:nr){
```

```
      u0[(iz-1)*nr+ir]=u10;
    }
    }
    ncall=0;
#
# ODE integration
  out=lsodes(y=u0,times=tout,func=pde2b,
      sparsetype ="sparseint",rtol=1e-6,
      atol=1e-6,maxord=5);
  nrow(out)
  ncol(out)
#
# Arrays for plotting numerical solution
  u1=matrix(0,nrow=nz,ncol=nout);
  for(it in 1:nout){
    cat(sprintf("\n"));
    ir=1;
    for(iz in 1:nz){
      u1[iz,it]=out[it,(iz-1)*nr+ir+1];
      u1[1,it]=u1b;
      cat(sprintf("%7.1f%7.1f%12.3e\n",
          it,(iz-1)*nr+ir+1,u1[iz,it]));
      }
  u1[1,it]=u1b
    }
#
# Calls to ODE routine
  cat(sprintf("\n\n ncall = %5d\n\n",ncall));
#
# Plot PDE solutions
#
# u1(z,t)
# 2D
  par(mfrow=c(1,1));
  matplot(x=z[2:nz],y=u1[2:nz,],type="l",xlab="z",ylab="u1(r=0,z,t)",
    lty=1,main="",lwd=2,col="black");
#
# 3D
  persp(z,tout,u1,theta=45,phi=30,xlim=c(zl,zu),ylim=c(t0,tf),
    xlab="z",ylab="t",zlab="u1(r=0,z,t)");
```

Listing 3.3: Main program for eqs. (1.2), Dirichlet BC.

We can note the following details about Listing 3.3 (with some repetition of the discussion of Listing 3.1 so that the following explanation is self contained).

- Previous workspaces are deleted.

```
#
# 2D TC model
#
# Delete previous workspaces
  rm(list=ls(all=TRUE))
```

- The R ODE integrator library deSolve is accessed [1].

```
#
# Access ODE integrator
  library("deSolve");
#
# Access functions for numerical solution
  setwd("f:/tissue engineering/chap3");
  source("pde2b.R");
```

Then the directory with the files for the solution of eqs. (1.2) is designated. Note that setwd (set working directory) uses / rather than the usual \. The ordinary differential equation/method of lines (ODE/MOL) routine, pde2b, is accessed through setwd.
- The model parameters are specified numerically.

```
#
# Parameters
  D1r=0.1;
  D1z=0.1;
  p=1;
  Qm=1;
  cm=1;
  pQm=p*Qm;
  u1b=1;
  u10=0;
```

In particular,
- u1b: Boundary value $u_1(r = 0, z = 0, t) = u_{1b} = 1$ in BC (1.2-5).
- u10: Initial value $u_1(r, z, t = 0) = u_{10} = 0$ in IC (1.2-2).
- A spatial grid in r for eq. (1.2-1) is defined with 11 points so that r = $0, 1/10=0.1, \ldots, 1$. The stem cell length is a normalized value, $r = r_u = 1$.

```
#
# Spatial grid (in r)
  nr=11;rl=0;ru=1;dr=(ru-rl)/(nr-1);dr2=dr^2;
  r=seq(from=rl,to=ru,by=dr);
```

- A spatial grid in z for eq. (1.2-1) is defined with 21 points so that $z = 0,1/20=0.05,\ldots,1$. The stem cell length is a normalized value, $z = z_u = 1$.

```
#
# Spatial grid (in z)
  nz=21;zl=0;zu=1;dz=(zu-zl)/(nz-1);dz2=dz^2;
  z=seq(from=zl,to=zu,by=dz);
```

- An interval in t is defined for 11 output points, so that tout=0,1/10=0.1,\ldots,1. The time scale is normalized with $t_f = 1$ specified as the final time that is considered appropriate, e.g., minutes, days.

```
#
# Independent variable for ODE integration
  t0=0;tf=1;nout=11;
  tout=seq(from=t0,to=tf,by=(tf-t0)/(nout-1));
```

- IC (1.2-2) is implemented.

```
#
# Initial condition (t=0)
  u0=rep(0,nr*nz);
  for(iz in 1:nz){
  for(ir in 1:nr){
    u0[(iz-1)*nr+ir]=u10;
  }
  }
    ncall=0;
```

Also, the counter for the calls to pde2b is initialized.
- The system of nr*nz=11*21=231 ODEs is integrated by the library integrator lsodes (available in deSolve, [1]). As expected, the inputs to lsodes are the ODE function, pde2b, the IC vector u0, and the vector of output values of t, tout. The length of u0 (231) informs lsodes how many ODEs are to be integrated. func,y,times are reserved names.

```
#
# ODE integration
  out=lsodes(y=u0,times=tout,func=pde2b,
```

```
      sparsetype ="sparseint",rtol=1e-6,
      atol=1e-6,maxord=5);
   nrow(out)
   ncol(out)
```

nrow,ncol confirm the dimensions of out.

- $u_1(r = 0, z, t)$ is placed in a matrix for plotting.

```
#
# Array for plotting numerical solution
  u1=matrix(0,nrow=nz,ncol=nout);
  for(it in 1:nout){
    cat(sprintf("\n      t        z      (iz-1)nr+ir  u1(r=0,z,t)\n"));
    ir=1;
    for(iz in 1:nz){
      u1[iz,it]=out[it,(iz-1)*nr+ir+1];
      cat(sprintf("%7.1f%9.3f%9.1f%16.3e\n",
          tout[it],z[iz],(iz-1)*nr+ir,u1[iz,it]));
    }
    u1[1,it]=u1b
  }
```

The offset +1 is required since the first element of the solution vectors in out is the value of t and the 2 to 232 elements are the 231 values of u_1. These dimensions from the preceding calls to nrow,ncol are confirmed in the subsequent output.

The boundary value $u_1(r = 0, z = 0, t)$ is set by an algebraic equation in pde2b, but lsodes returns only solutions to ODEs. So the boundary value is set in the main program above as u1[1,it]=u1b.

- The number of calls to pde2b is displayed at the end of the solution.

```
#
# Calls to ODE routine
  cat(sprintf("\n\n ncall = %5d\n\n",ncall));
```

- $u_1(r = 0, z, t)$ is plotted in 2D against z and parametrically in t with the R utility matplot, and in 3D with the R utility persp. par(mfrow=c(1,1)) specifies a 1×1 matrix of plots, that is, one plot on a page.

```
#
# Plot PDE solutions
#
# u1(z,t)
# 2D
  par(mfrow=c(1,1));
```

```
    matplot(x=z[2:nz],y=u1[2:nz,],type="l",xlab="z",ylab="u1(r=0,z,t)",
      lty=1,main="",lwd=2,col="black");
#
# 3D
    persp(z,tout,u1,theta=45,phi=30,xlim=c(zl,zu),ylim=c(t0,tf),
      xlab="z",ylab="t",zlab="u1(r=0,z,t)");
```

By using [2:nz], the discontinuity between the IC $u_1(r = 0, z, t = 0) = 0$ and the BC $u_1(r = 0, z = 0, t) = 1$ is not included in the 2D plot (Fig. 3.3-1). Including the discontinuity is left as an exercise.

This completes the discussion of the main program for eqs. (1.2). The ODE/MOL routine pde2b called by lsodes from the main program of Listing 3.3 for the numerical MOL integration of eqs. (1.2) is next.

3.1.5 ODE/MOL routine

The ODE/MOL routine, pde2b, called by lsodes from the main program of Listing 3.3 follows.

```
  pde2b=function(t,u,parm){
#
# Function pde2b computes the t derivative
# of u1(r,z,t)
#
# 1D to 2D vector
  u1=matrix(0,nrow=nr,ncol=nz);
  for(iz in 1:nz){
  for(ir in 1:nr){
    u1[ir,iz]=u[(iz-1)*nr+ir];
  }
  }
#
# u1rr, Neumann BC, r=0,ru
  u1rr=matrix(0,nrow=nr,ncol=nz);
  for(iz in 1:nz){
  for(ir in 1:nr){
    if(ir==1){
      u1rr[1,iz]=4*(u1[2,iz]-u1[1,iz])/dr2;}
    if(ir==nr){
      u1rr[nr,iz]=2*(u1[nr-1,iz]-u1[nr,iz])/dr2;}
    if((ir>1)&(ir<nr)){
```

```
      u1rr[ir,iz]=(u1[ir+1,iz]-2*u1[ir,iz]+u1[ir-1,iz])/dr2+
              (1/r[ir])*(u1[ir+1,iz]-u1[ir-1,iz])/(2*dr);}
  }
  }
#
# u1zz, Dirichlet BC, z=zl Neumann BC, z=zu
  u1zz=matrix(0,nrow=nr,ncol=nz);
  for(iz in 1:nz){
  for(ir in 1:nr){
    if(iz==1){
      u1[ir,1]=u1b;}
    if(iz==nz){
      u1zz[ir,nz]=2*(u1[ir,nz-1]-u1[ir,nz])/dz2;}
    if((iz>1)&(iz<nz)){
      u1zz[ir,iz]=(u1[ir,iz+1]-2*u1[ir,iz]+u1[ir,iz-1])/dz2;}
  }
  }
#
# u1t
  u1t=matrix(0,nrow=nr,ncol=nz);
  for(iz in 2:nz){
  for(ir in 1:nr){
#
#   Michaelis-Menten
    q=pQm*u1[ir,iz]/(cm+u1[ir,iz]);
#
#   PDE
    u1t[ir,iz]=D1r*u1rr[ir,iz]+D1z*u1zz[ir,iz]-q;
    u1t[ir,1]=0;
  }
  }
#
# 2D to 1D vector
  ut=rep(0,nr*nz);
  for(iz in 1:nz){
  for(ir in 1:nr){
    ut[(iz-1)*nr+ir]=u1t[ir,iz];
  }
  }
#
# Increment calls to pde2b
  ncall <<- ncall+1;
```

```
#
# Return derivative vector
  return(list(c(ut)));
  }
```

Listing 3.4: ODE/MOL routine for eqs. (1.2), Dirichlet BC.

We can note the following details about Listing 3.4 (with some repetition of the discussion of Listing 3.2 so that the following explanation is self contained).

- The function is defined.

```
    pde2b=function(t,u,parm){
#
# Function pde2b computes the t derivative
# of u1(r,z,t)
```

   t is the current value of t in eqs. (1.2). u is the 231-vector of ODE/PDE dependent variables. parm is an argument to pass parameters to pde2b (unused, but required in the argument list). The arguments must be listed in the order stated to properly interface with lsodes called in the main program of Listing 3.3. The derivative vector of the LHS of eq. (1.2-1) is calculated and returned to lsodes as explained subsequently.
- The input vector u is placed in a matrix u1 to facilitate the programming of eq. (1.2-1).

```
#
# 1D vector to 2D matrix
  u1=matrix(0,nrow=nr,ncol=nz);
  for(iz in 1:nz){
  for(ir in 1:nr){
    u1[ir,iz]=u[(iz-1)*nr+ir];
  }
  }
```

- The radial group in eq. (1.2-1), $\dfrac{\partial^2 u_1}{\partial^2 r} + \dfrac{1}{r}\dfrac{\partial u_1}{\partial r}$, is programmed (a detailed explanation of this code follows after Listing 3.2).

```
#
# u1rr, Neumann BC, r=0,ru
  u1rr=matrix(0,nrow=nr,ncol=nz);
  for(iz in 1:nz){
  for(ir in 1:nr){
    if(ir==1){
      u1rr[1,iz]=4*(u1[2,iz]-u1[1,iz])/dr2;}
    if(ir==nr){
```

```
                u1rr[nr,iz]=2*(u1[nr-1,iz]-u1[nr,iz])/dr2;}
          if((ir>1)&(ir<nr)){
                u1rr[ir,iz]=(u1[ir+1,iz]-2*u1[ir,iz]+u1[ir-1,iz])/dr2+
                          (1/r[ir])*(u1[ir+1,iz]-u1[ir-1,iz])/(2*dr);}
    }
    }
```

- The derivative $\frac{\partial^2 u_1}{\partial z^2}$ in eq. (1.2-1) is programmed.

```
#
# u1zz, Dirichlet BC, z=zl Neumann BC, z=zu
  u1zz=matrix(0,nrow=nr,ncol=nz);
  for(iz in 1:nz){
  for(ir in 1:nr){
    if(iz==1){
      u1[ir,1]=u1b;}
    if(iz==nz){
      u1zz[ir,nz]=2*(u1[ir,nz-1]-u1[ir,nz])/dz2;}
    if((iz>1)&(iz<nz)){
      u1zz[ir,iz]=(u1[ir,iz+1]-2*u1[ir,iz]+u1[ir,iz-1])/dz2;}
  }
  }
```

This coding requires some additional explanation.

- The derivative $\frac{\partial^2 u_1}{\partial z^2}$ in eq. (1.2-1) is approximated with a three point centered (in z) finite difference (FD).

$$\frac{\partial^2 u_1(r,z,t)}{\partial z^2} \approx \frac{(u_1(r,z+\Delta z,t) - 2u_1(r,z,t) + u_1(r,z-\Delta z,t))}{\Delta z^2} + O(\Delta z^2) \quad (3.1\text{-}1)$$

$O(\Delta z^2)$ indicates that the error in the FD approximation is second order in Δz. The variation in the numerical solution of eq. (1.2-1) can be studied as a function of the FD increment Δz by varying nz in Listing 3.3.

- Dirichlet BC (1.2-5) is applied at $z = z_l = 0$.

$$u_1(r, z = 0, t) = u_{1b} \quad (3.1\text{-}2)$$

and is programmed as

```
        if(iz==1){
          u1[ir,1]=u1b;}
```

- At $z = z_u$, Neumann BC (1.2-6) is used to eliminate the fictitious value $u_1(r, z_u + \Delta z, t)$, and eq. (3.1-1) is

$$\frac{\partial^2 u_1(z_u, t)}{\partial z^2} \approx \frac{2(u_1(z_u - \Delta z, t) - u_1(z_u, t))}{\Delta z^2} \quad (3.1\text{-}3)$$

that is programmed as

```
if(iz==nz){
  u1zz[ir,nz]=2*(u1[ir,nz-1]-u1[ir,nz])/dz2;}
```

• For the interior points $z_l < z < z_u$, eq. (3.1-1) is programmed as

```
if((iz>1)&(iz<nz)){
  u1zz[ir,iz]=(u1[ir,iz+1]-2*u1[ir,iz]+u1[ir,iz-1])/dz2;}
  }
  }
```

The final } concludes the for in z.

$\dfrac{\partial^2 u_1}{\partial z^2}$ is now available for use in the programming of eq. (1.2-1).

• Eq. (1.2-1) is programmed in the MOL format.

```
#
# u1t
  u1t=matrix(0,nrow=nr,ncol=nz);
  for(iz in 2:nz){
  for(ir in 1:nr){
#
#    Michaelis-Menten
     q=pQm*u1[ir,iz]/(cm+u1[ir,iz]);
#
#    PDE
     u1t[ir,iz]=D1r*u1rr[ir,iz]+D1z*u1zz[ir,iz]-q;
     u1t[ir,1]=0;
  }
  }
```

$\dfrac{\partial u_1(r, z = z_l, t)}{\partial t} = 0$ so that $u_1(z = z_l, t)$ remains at the boundary value (eq. (1.2-5)).

• The 231 ODE derivatives are placed in the vector ut for return to lsodes to take the next step in t along the solution.

```
#
# 2D matrix to 1D vector
  ut=rep(0,nr*nz);
  for(iz in 1:nz){
  for(ir in 1:nr){
    ut[(iz-1)*nr+ir]=u1t[ir,iz];
  }
  }
```

- The counter for the calls to `pde2b` is incremented and returned to the main program of Listing 3.3 by `<<-`.

```
#
# Increment calls to pde2b
  ncall <<- ncall+1;
```

- The vector `ut` is returned as a `list` as required by `lsodes`. `c` is the R vector utility.

```
#
# Return derivative vector
  return(list(c(ut)));
  }
```

The final `}` concludes `pde2b`.

This completes the discussion of `pde2b`.

3.1.6 Numerical, graphical output

Abbreviated numerical output from the main program of Listing 3.3 and ODE/MOL routine `pde2b` of Listing 3.4 follows.

Table 3.3 Numerical output from Listings 3.3, 3.4, Dirichlet BC.

```
[1]  11

[1]  232

        t        z      (iz-1)nr+ir   u1(r=0,z,t)
       0.0    0.000        1.0         0.000e+00
       0.0    0.050       12.0         0.000e+00
       0.0    0.100       23.0         0.000e+00
                              .                .
                              .                .
                              .                .
           Output for (iz-1)nr+ir = 34,...
                      188 removed
                              .                .
                              .                .
                              .                .
       0.0    0.900      199.0         0.000e+00
       0.0    0.950      210.0         0.000e+00
       0.0    1.000      221.0         0.000e+00
```

continued on next page

Table 3.3   (continued)

t	z	(iz-1)nr+ir	u1(r=0,z,t)
0.1	0.000	1.0	0.000e+00
0.1	0.050	12.0	7.116e-01
0.1	0.100	23.0	4.651e-01
.	.		.
.	.		.
.	.		.

Output for (iz-1)nr+ir = 34,...
188 removed

	.		.
	.		.
	.		.

0.1	0.900	199.0	1.132e-08
0.1	0.950	210.0	2.330e-09
0.1	1.000	221.0	8.579e-10
.	.		.
.	.		.
.	.		.

Output for t = 0.2,...0.9 removed

	.		.
	.		.
	.		.

t	z	(iz-1)nr+ir	u1(r=0,z,t)
1.0	0.000	1.0	0.000e+00
1.0	0.050	12.0	8.682e-01
1.0	0.100	23.0	7.487e-01
.	.		.
.	.		.
.	.		.

Output for (iz-1)nr+ir = 34,...
188 removed

	.		.
	.		.
	.		.

1.0	0.900	199.0	3.055e-02
1.0	0.950	210.0	2.723e-02
1.0	1.000	221.0	2.614e-02

ncall =   364

We can note the following details about this output.

- 11 *t* output points as the first dimension of the solution matrix out from lsodes as programmed in the main program of Listing 3.3 (with nout=11).

- The solution matrix out returned by lsodes has 232 elements as a second dimension. The first element is the value of t. Elements 2 to 232 are $u_1(r, z, t)$ from eqs. (1.2) (for the spatial grid with $n_r = 11, n_z = 21, (n_r)(n_z) = (11)(21) = 231$ points).
- The solution is displayed for t=0,1/10=0.1,...,1 as programmed in Listing 3.3.
- The solution $u_1(r = 0, z, t)$ is displayed for the index (iz-1)*nr+ir=1,12,...,221 with iz=1,...,21, ir=1 programmed in Listing 3.3.
- IC (1.2-2) is confirmed (at $t = 0$).
  Note in particular the value of $u_1(r = 0, z = 0, t = 0) = g_2(r - 0, t) = 0$ which reflects BC (1.2-5) programmed in Listing 3.4.
- The computational effort as indicated by ncall = 364 is modest so that lsodes computed the solution to eqs. (1.2) efficiently.

The graphical output is in Figs. 3.3.

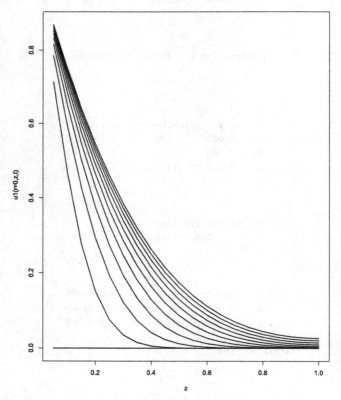

FIGURE 3.3-1 $u_1(r = 0, z, t)$ from eqs. (1.2), 2D, Dirichlet BC.

This solution displays the properties discussed for Table 3.3 (BC (1.2-5) that moves the solution from IC (1.2-2)).

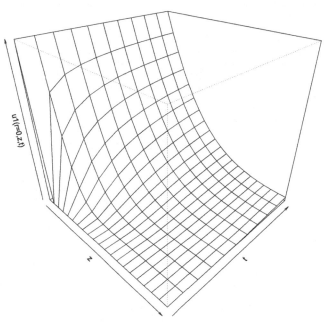

FIGURE 3.3-2 $u_1(r = 0, z, t)$ from eqs. (1.2), 3D, Dirichlet BC.

This concludes the discussion of Listings 3.3, 3.4 for which a Dirichlet BC is specified at $z = z_l$ and a Neumann BC at $z = z_u$.

3.2 Summary and conclusions

A 2D O_2 diffusion model is implemented in this chapter for: (1) two Neumann BCs, (2) one Dirichlet and one Neumann BC applied to the spatial domain of the stem cells seeded on a scaffold. The R routines are based on the MOL with the spatial derivatives approximated by finite differences. For (2), the increase of O_2 concentration provides an essential condition for the stem cell differentiation that could be the basis for tissue engineering and regenerative medicine.

The extension to a model with two PDEs for stem cell O_2 and nutrient concentration is considered next.

Reference

[1] K. Soetaert, J. Cash, F. Mazzia, Solving Differential Equations in R, Springer-Verlag, Heidelberg, Germany, 2012.

Two PDE model formulation

Introduction

The 2D PDE model of eqs. (1.2) in Chapter 1 is implemented in this chapter for two simultaneous PDEs.

4.1 2D two PDE model implementation

Eqs. (1.2) are extended to a 2D two PDE model for the stem cell concentration of (1) O_2, $u_1(r, z, t)$ and (2) a nutrient $u_2(r, z, t)$. The model is stated next.

$$\frac{\partial u_1}{\partial t} = D_{1,r} \left(\frac{\partial^2 u_1}{\partial r^2} + \frac{1}{r}\frac{\partial u_1}{\partial r} \right) + D_{1,z}\frac{\partial^2 u_1}{\partial z^2} - q_1 \tag{4.1-1}$$

$D_{1,r}, D_{1,z}$ are effective diffusivities in r and z in the tissue construct (TC).

Eq. (4.1-1) is first order in t and requires one IC.

$$u_1(r, z, t = 0) = f_1(r, z) \tag{4.1-2}$$

$f_1(r, z)$ is a function to be specified.

Eq. (4.1-1) is second order in r and z, and requires two BC in each of these spatial independent variables. For r, the BCs are homogeneous Neumann (no flux).

$$\frac{\partial u_1(r = 0, z, t)}{\partial r} = 0 \tag{4.1-3}$$

$$\frac{\partial u_1(r = r_u, z, t)}{\partial r} = 0 \tag{4.1-4}$$

r_u is the outer boundary in r. BC (4.1-3) specifies symmetry at $r = 0$. BC (4.1-4) specifies zero flux at $r = r_u$.

For z, the BCs are Dirichlet at $z = z_l = 0$ and homogeneous Neumann at $z = z_u = 1$.

$$u_1(r, z = 0, t) = g_1(r, t) \tag{4.1-5}$$

$$\frac{\partial u_1(r, z = z_u, t)}{\partial z} = 0 \tag{4.1-6}$$

$g_1(r, t)$ is a function to be specified.

$q_1 = q_1(u_1(r, z, t), u_2(r, t))$ is a two component Michaelis-Menten rate.

$$q_1 = pQm_1\frac{u_1}{cm_1 + u_1}u_2 \tag{4.1-7}$$

The analogous equations for $u_2(r, z, t)$ follow.

$$\frac{\partial u_2}{\partial t} = D_{2,r}\left(\frac{\partial^2 u_2}{\partial r^2} + \frac{1}{r}\frac{\partial u_2}{\partial r}\right) + D_{2,z}\frac{\partial^2 u_2}{\partial z^2} - q_2 \qquad (4.2\text{-}1)$$

Eq. (4.2-1) is first order in t and requires one IC.

$$u_2(r, z, t = 0) = f_2(r, z) \qquad (4.2\text{-}2)$$

$f_2(r, z)$ is a function to be specified.

Eq. (4.2-1) is second order in r and z, and requires two BC in each of these spatial independent variables. For r, the BCs are homogeneous Neumann (no flux).

$$\frac{\partial u_2(r = 0, z, t)}{\partial r} = 0 \qquad (4.2\text{-}3)$$

$$\frac{\partial u_2(r = r_u, z, t)}{\partial r} = 0 \qquad (4.2\text{-}4)$$

BC (4.2-3) specifies symmetry at $r = 0$. BC (4.2-4) specifies zero flux at $r = r_u$.

For z, the BCs are Dirichlet at $z = z_l = 0$ and homogeneous Neumann at $z = z_u = 1$.

$$u_2(r, z = 0, t) = g_2(r, t) \qquad (4.2\text{-}5)$$

$$\frac{\partial u_2(r, z = z_u, t)}{\partial z} = 0 \qquad (4.2\text{-}6)$$

$g_2(r, t)$ is a function to be specified.

$q_2 = q_2(u_1(r, z, t), u_2(r, t))$ is a two component Michaelis-Menten rate.

$$q_2 = pQm_2\frac{u_2}{cm_2 + u_2}u_1 \qquad (4.2\text{-}7)$$

q_1, q_2 couple the PDEs, eqs. (4.1-1), (4.2-1), since these are functions of both $u_1(r, z, t)$ and $u_2(r, z, t)$. In other words, eqs. (4.1-1) and (4.2-1) must be integrated (solved) simultaneously and not sequentially (one at a time).

Also, q_1, q_2 are nonlinear, but the integration of eqs. (4.1-1), (4.2-1) is straightforward numerically (as demonstrated in ODE/MOL function pde2a discussed subsequently), while analytical solution would be difficult.

Physically, q_1, q_2 are volumetric rates of O_2 consumption (represented with $u_1(r, z, t)$) and nutrient consumption (represented with $u_2(r, z, t)$), i.e., metabolism that energizes the stem cells. Alternative forms of these functions can be considered, for example,

$$q_1 = pQm_1\frac{u_1}{cm_1 + u_1}pQm_2\frac{u_2}{cm_2 + u_2} \qquad (4.2\text{-}8)$$

$$q_2 = pQm_2\frac{u_2}{cm_2 + u_2}pQm_1\frac{u_1}{cm_1 + u_1} \qquad (4.2\text{-}9)$$

Implementation of these alternative Michaelis-Menten rates (in pde2a) is left as an exercise.

4.1.1 Main program, test cases

A main program for eqs. (4.1), (4.2) follows.

```
#
# 2D TC model
#
# Delete previous workspaces
  rm(list=ls(all=TRUE))
#
# Access ODE integrator
  library("deSolve");
#
# Access functions for numerical solution
  setwd("f:/tissue engineering/chap4");
  source("pde2a.R");
#
# Parameters
#
# Component 1
  D1r=0.1;
  D1z=0.1;
  p1=1;
  Qm1=1;
  cm1=1;
  pQm1=p1*Qm1;
  u10=1;
#
# Component 2
  D2r=0.1;
  D2z=0.1;
  p2=1;
  Qm2=1;
  cm2=1;
  pQm2=p2*Qm2;
  u20=1;
  ncase=1;
#
# Spatial grid (in r)
  nr=11;rl=0;ru=1;dr=(ru-rl)/(nr-1);dr2=dr^2;
  r=seq(from=rl,to=ru,by=dr);
#
```

```
# Spatial grid (in z)
  nz=21;zl=0;zu=1;dz=(zu-zl)/(nz-1);dz2=dz^2;
  z=seq(from=zl,to=zu,by=dz);
#
# Independent variable for ODE integration
  t0=0;tf=1;nout=11;
  tout=seq(from=t0,to=tf,by=(tf-t0)/(nout-1));
#
# Initial condition (t=0)
  u0=rep(0,2*nr*nz);
  if(ncase==1){
    for(iz in 1:nz){
    for(ir in 1:nr){
      u0[(iz-1)*nr+ir]      =u10;
      u0[(iz-1)*nr+ir+nr*nz]=u20;
    }
    }
  }
  if(ncase==2){
    for(iz in 1:nz){
    for(ir in 1:nr){
      u0[(iz-1)*nr+ir]=
        sin(pi*(ir-1)/(nr-1))*
        sin(pi*(iz-1)/(nz-1));
      u0[(iz-1)*nr+ir+nr*nz]=
        sin(pi*(ir-1)/(nr-1))*
        sin(pi*(iz-1)/(nz-1));
    }
    }
  }
  ncall=0;
#
# ODE integration
  out=lsodes(y=u0,times=tout,func=pde2a,
      sparsetype ="sparseint",rtol=1e-6,
      atol=1e-6,maxord=5);
  nrow(out)
  ncol(out)
#
# Arrays for plotting numerical solution
  u1=matrix(0,nrow=nz,ncol=nout);
  u2=matrix(0,nrow=nz,ncol=nout);
```

```
    for(it in 1:nout){
      cat(sprintf("\n       t        z      (iz-1)nr+ir  u1(r=0,z,t)\n"));
      ir=1;
      for(iz in 1:nz){
        u1[iz,it]=out[it,(iz-1)*nr+ir+1];
        cat(sprintf("%7.1f%9.3f%9.1f%16.3e\n",
            tout[it],z[iz],(iz-1)*nr+ir,u1[iz,it]));
       }
      cat(sprintf("\n       t        z      (iz-1)nr+ir  u2(r=0,z,t)\n"));
      for(iz in 1:nz){
        u2[iz,it]=out[it,(iz-1)*nr+ir+1+nr*nz];
        cat(sprintf("%7.1f%9.3f%9.1f%16.3e\n",
            tout[it],z[iz],(iz-1)*nr+ir,u2[iz,it]));
      }
  }
#
# Calls to ODE routine
  cat(sprintf("\n\n ncall = %5d\n\n",ncall));
#
# Plot PDE solutions
#
# u1(r,z,t)
# 2D
  par(mfrow=c(1,1));
  matplot(x=z,y=u1,type="l",xlab="z",ylab="u1(r=0,z,t)",
    lty=1,main="",lwd=2,col="black");
#
# 3D
  persp(z,tout,u1,theta=45,phi=30,xlim=c(zl,zu),ylim=c(t0,tf),
    xlab="z",ylab="t",zlab="u1(r=0,z,t)");
#
# u2(r,z,t)
# 2D
  par(mfrow=c(1,1));
  matplot(x=z,y=u2,type="l",xlab="z",ylab="u2(r=0,z,t)",
    lty=1,main="",lwd=2,col="black");
#
# 3D
  persp(z,tout,u2,theta=45,phi=30,xlim=c(zl,zu),ylim=c(t0,tf),
    xlab="z",ylab="t",zlab="u2(r=0,z,t)");
```

Listing 4.1: Main program for eqs. (4.1), (4.2).

We can note the following details about Listing 4.1 (with some repetition of the discussion of Listings 2.1, 3.1 so that the following explanation is self contained).

- Previous workspaces are deleted.

```
#
# 2D TC model
#
# Delete previous workspaces
  rm(list=ls(all=TRUE))
```

- The R ODE integrator library deSolve is accessed [1].

```
#
# Access ODE integrator
  library("deSolve");
#
# Access functions for numerical solution
  setwd("f:/tissue engineering/chap4");
  source("pde2a.R");
```

Then the directory with the files for the solution of eqs. (4.1), (4.2) is designated. Note that setwd (set working directory) uses / rather than the usual \. The ordinary differential equation/method of lines (ODE/MOL)[1] routine, pde2a.R, is accessed through setwd.

- The model parameters are specified numerically.

```
#
# Parameters
#
# Component 1
  D1r=0.1;
  D1z=0.1;
  p1=1;
  Qm1=1;
  cm1=1;
  pQm1=p1*Qm1;
  u10=1;
#
# Component 2
  D2r=0.1;
```

[1] The method of lines is a general numerical algorithm for PDEs in which the boundary value (spatial) derivatives are replaced with algebraic approximations, in this case, finite differences (FDs). The resulting system of initial value ODEs is then integrated (solved) with a library ODE integrator.

```
   D2z=0.1;
   p2=1;
   Qm2=1;
   cm2=1;
   pQm2=p2*Qm2;
   u20=1;
   ncase=1;
```

where

- D1r, D1z, D2r, D2z: Effective O_2 diffusivities in eqs. (4.1-1), (4.2-1).
- p1,Qm1,cm1: Parameters in q_1 of eq. (4.1-7).
- p2,Qm2,cm2: Parameters in q_2 of eq. (4.2-7).
- u10: Initial condition (IC) for eq. (4.1-1) (eq. (4.1-2)).
- u20: Initial condition (IC) for eq. (4.2-1) (eq. (4.2-2)).
- ncase: Index for selection of the initial condition functions of eq. (4.1-1), (4.2-1) ($f_1(r, z)$, $f_2(r, z)$ in eqs. (4.1-2), (4.2-2)).

- A spatial grid in r for eqs. (4.1-1) (4.2-1) is defined with 11 points so that r = 0,1/10= 0.1, ..., 1. The stem cell radius is a normalized value, $r = r_u = 1$.

```
#
# Spatial grid (in r)
   nr=11;rl=0;ru=1;dr=(ru-rl)/(nr-1);dr2=dr^2;
   r=seq(from=rl,to=ru,by=dr);
```

- A spatial grid in z for eqs. (4.1-1), (4.2-1) is defined with 21 points so that z = 0,1/20=0.05, ..., 1. The stem cell length is a normalized value, $z = z_u = 1$.

```
#
# Spatial grid (in z)
   nz=21;zl=0;zu=1;dz=(zu-zl)/(nz-1);dz2=dz^2;
   z=seq(from=zl,to=zu,by=dz);
```

- An interval in t is defined for 11 output points, so that tout=0,1/10=0.1, ..., 1. The time scale is normalized with $t_f = 1$ specified as the final time that is considered appropriate, e.g., minutes, days.

```
#
# Independent variable for ODE integration
   t0=0;tf=1;nout=11;
   tout=seq(from=t0,to=tf,by=(tf-t0)/(nout-1));
```

- ICs (4.1-2), (4.2-2) are implemented for ncase=1,2.

```
#
# Initial condition (t=0)
```

```
    u0=rep(0,2*nr*nz);
    if(ncase==1){
      for(iz in 1:nz){
      for(ir in 1:nr){
        u0[(iz-1)*nr+ir]       =u10;
        u0[(iz-1)*nr+ir+nr*nz]=u20;
      }
      }
    }
    if(ncase==2){
      for(iz in 1:nz){
      for(ir in 1:nr){
        u0[(iz-1)*nr+ir]=
          sin(pi*(ir-1)/(nr-1))*
          sin(pi*(iz-1)/(nz-1));
        u0[(iz-1)*nr+ir+nr*nz]=
          sin(pi*(ir-1)/(nr-1))*
          sin(pi*(iz-1)/(nz-1));
      }
      }
    }
    ncall=0;
```

For ncase=1, the IC functions are constant, u10, u20. For ncase=2, the IC functions are sine functions (discussed subsequently).

The index (iz-1)*nr+ir is for eq. (4.1-1), and the index (iz-1)*nr+ir+nr*nz is for eq. (4.2-1).

Also, the counter for the calls to pde2a is initialized.

• The system of 2*nr*nz=2*(11*21)=462 ODEs is integrated by the library integrator lsodes (available in deSolve, [1]). As expected, the inputs to lsodes are the ODE function, pde2a, the IC vector u0, and the vector of output values of t, tout. The length of u0 (462) informs lsodes how many ODEs are to be integrated. func,y,times are reserved names.

```
#
# ODE integration
  out=lsodes(y=u0,times=tout,func=pde2a,
      sparsetype ="sparseint",rtol=1e-6,
      atol=1e-6,maxord=5);
  nrow(out)
  ncol(out)
```

nrow,ncol confirm the dimensions of out.

- $u_1(r, z, t)$, $u_2(r, z, t)$ are placed in matrices for subsequent plotting.

```
#
# Arrays for plotting numerical solution
  u1=matrix(0,nrow=nz,ncol=nout);
  u2=matrix(0,nrow=nz,ncol=nout);
  for(it in 1:nout){
    cat(sprintf("\n      t       z      (iz-1)nr+ir  u1(r=0,z,t)\n"));
    ir=1;
    for(iz in 1:nz){
      u1[iz,it]=out[it,(iz-1)*nr+ir+1];
      cat(sprintf("%7.1f%9.3f%9.1f%16.3e\n",
          tout[it],z[iz],(iz-1)*nr+ir,u1[iz,it]));
    }
    cat(sprintf("\n      t       z      (iz-1)nr+ir  u2(r=0,z,t)\n"));
    for(iz in 1:nz){
      u2[iz,it]=out[it,(iz-1)*nr+ir+1+nr*nz];
      cat(sprintf("%7.1f%9.3f%9.1f%16.3e\n",
          tout[it],z[iz],(iz-1)*nr+ir,u2[iz,it]));
    }
  }
```

Since $u_1(r, z, t)$, $u_2(r, z, t)$ are functions of three variables, r, z, t, they cannot be plotted in 3D. Therefore, the solution at $r = 0$ is selected with ir=1, that is, $u_1(r = 0, z, t)$, $u_2(r = 0, z, t)$ are plotted.

The offset +1 is required because the first element of the solution vectors in out is the value of t and the 2 to 463 elements are the $2(11)(21) = 462$ values of u_1, u_2. These dimensions from the preceding calls to nrow,ncol are confirmed in the subsequent output.

- The number of calls to pde2a is displayed at the end of the solution.

```
#
# Calls to ODE routine
  cat(sprintf("\n\n ncall = %5d\n\n",ncall));
```

- par(mfrow=c(1,1)) specifies a 1×1 matrix of plots, that is, one plot on a page.
  $u_1(r = 0, z, t)$, $u_2(r = 0, z, t)$ are plotted in 2D against z and parametrically in t with the R utility matplot, and in 3D against z and t with the R utility persp.

```
#
# Plot PDE solutions
#
# u1(r,z,t)
# 2D
```

```
    par(mfrow=c(1,1));
    matplot(x=z,y=u1,type="l",xlab="z",ylab="u1(r=0,z,t)",
      lty=1,main="",lwd=2,col="black");
  #
  # 3D
    persp(z,tout,u1,theta=45,phi=30,xlim=c(zl,zu),ylim=c(t0,tf),
      xlab="z",ylab="t",zlab="u1(r=0,z,t)");
  #
  # u2(r,z,t)
  # 2D
    par(mfrow=c(1,1));
    matplot(x=z,y=u2,type="l",xlab="z",ylab="u2(r=0,z,t)",
      lty=1,main="",lwd=2,col="black");
  #
  # 3D
    persp(z,tout,u2,theta=45,phi=30,xlim=c(zl,zu),ylim=c(t0,tf),
      xlab="z",ylab="t",zlab="u2(r=0,z,t)");
```

This completes the discussion of the main program for eqs. (4.1-1), (4.2-1). The ODE/MOL routine pde2a called by lsodes from the main program for the numerical MOL integration of eqs. (4.1), (4.2) is next.

4.1.2 ODE/MOL routine

pde2a called in the main program of Listing 4.1 follows.

```
  pde2a=function(t,u,parm){
#
# Function pde2a computes the t derivative
# of u1(r,z,t), u2(r,z,t)
#
# 1D vector to 2D matrices
  u1=matrix(0,nrow=nr,ncol=nz);
  u2=matrix(0,nrow=nr,ncol=nz);
  for(iz in 1:nz){
  for(ir in 1:nr){
    u1[ir,iz]=u[(iz-1)*nr+ir];
    u2[ir,iz]=u[(iz-1)*nr+ir+nr*nz];
  }
  }
#
# u1rr, Neumann BC, r=0,ru
  u1rr=matrix(0,nrow=nr,ncol=nz);
```

```
    for(iz in 1:nz){
    for(ir in 1:nr){
      if(ir==1){
        u1rr[1,iz]=4*(u1[2,iz]-u1[1,iz])/dr2;}
      if(ir==nr){
        u1rr[nr,iz]=2*(u1[nr-1,iz]-u1[nr,iz])/dr2;}
      if((ir>1)&(ir<nr)){
        u1rr[ir,iz]=(u1[ir+1,iz]-2*u1[ir,iz]+u1[ir-1,iz])/dr2+
                    (1/r[ir])*(u1[ir+1,iz]-u1[ir-1,iz])/(2*dr);}
    }
    }
#
# u1zz, Neumann BC, z=0,zu
  u1zz=matrix(0,nrow=nr,ncol=nz);
    for(iz in 1:nz){
    for(ir in 1:nr){
      if(iz==1){
        u1zz[ir,1]=2*(u1[ir,2]-u1[ir,1])/dz2;}
      if(iz==nz){
        u1zz[ir,nz]=2*(u1[ir,nz-1]-u1[ir,nz])/dz2;}
      if((iz>1)&(iz<nz)){
        u1zz[ir,iz]=(u1[ir,iz+1]-2*u1[ir,iz]+u1[ir,iz-1])/dz2;}
    }
    }
#
# u2rr, Neumann BC, r=0,ru
  u2rr=matrix(0,nrow=nr,ncol=nz);
    for(iz in 1:nz){
    for(ir in 1:nr){
      if(ir==1){
        u2rr[1,iz]=4*(u2[2,iz]-u2[1,iz])/dr2;}
      if(ir==nr){
        u2rr[nr,iz]=2*(u2[nr-1,iz]-u2[nr,iz])/dr2;}
      if((ir>1)&(ir<nr)){
        u2rr[ir,iz]=(u2[ir+1,iz]-2*u2[ir,iz]+u2[ir-1,iz])/dr2+
                    (1/r[ir])*(u2[ir+1,iz]-u2[ir-1,iz])/(2*dr);}
    }
    }
#
# u2zz, Neumann BC, z=0,zu
  u2zz=matrix(0,nrow=nr,ncol=nz);
    for(iz in 1:nz){
```

```
    for(ir in 1:nr){
      if(iz==1){
        u2zz[ir,1]=2*(u2[ir,2]-u2[ir,1])/dz2;}
      if(iz==nz){
        u2zz[ir,nz]=2*(u2[ir,nz-1]-u2[ir,nz])/dz2;}
      if((iz>1)&(iz<nz)){
        u2zz[ir,iz]=(u2[ir,iz+1]-2*u2[ir,iz]+u2[ir,iz-1])/dz2;}
    }
  }
#
# u1t, u2t
  u1t=matrix(0,nrow=nr,ncol=nz);
  u2t=matrix(0,nrow=nr,ncol=nz);
  for(iz in 1:nz){
  for(ir in 1:nr){
#
#   Michaelis-Menten
    q1=pQm1*u1[ir,iz]/(cm1+u1[ir,iz])*u2[ir,iz];
    q2=pQm2*u2[ir,iz]/(cm2+u2[ir,iz])*u1[ir,iz];
#
#   PDE
    u1t[ir,iz]=D1r*u1rr[ir,iz]+D1z*u1zz[ir,iz]-q1;
    u2t[ir,iz]=D2r*u2rr[ir,iz]+D2z*u2zz[ir,iz]-q2;
  }
  }
#
# 2D matrices to 1D vector
  ut=rep(0,2*nr*nz)
  for(iz in 1:nz){
  for(ir in 1:nr){
    ut[(iz-1)*nr+ir]       =u1t[ir,iz];
    ut[(iz-1)*nr+ir+nr*nz]=u2t[ir,iz];
  }
  }
#
# Increment calls to pde2a
  ncall <<- ncall+1;
#
# Return derivative vector
  return(list(c(ut)));
  }
```

Listing 4.2: ODE/MOL routine for eqs. (4.1), (4.2).

We can note the following details about Listing 4.2 (with some repetition of the discussion of Listings 2.2, 3.2 so that the following explanation is self contained).

- The function is defined.

```
   pde2a=function(t,u,parm){
#
# Function pde2a computes the t derivative
# of u1(r,z,t), u2(r,z,t)
```

t is the current value of *t* in eqs. (4.1-1), (4.2-1). u is the 462-vector of ODE/PDE dependent variables. parm is an argument to pass parameters to pde2a (unused, but required in the argument list). The arguments must be listed in the order stated to properly interface with lsodes called in the main program of Listing 4.1. The derivative vector of the LHS of eqs. (4.1-1), (4.2-1) is calculated and returned to lsodes as explained subsequently.

- The input vector u is placed in matrices u1, u2 to facilitate the programming of eqs. (4.1-1), (4.2-1).

```
#
# 1D vector to 2D matrices
   u1=matrix(0,nrow=nr,ncol=nz);
   u2=matrix(0,nrow=nr,ncol=nz);
   for(iz in 1:nz){
   for(ir in 1:nr){
     u1[ir,iz]=u[(iz-1)*nr+ir];
     u2[ir,iz]=u[(iz-1)*nr+ir+nr*nz];
   }
   }
```

- The radial derivative group $\dfrac{\partial^2 u_1}{\partial r^2} + \dfrac{1}{r}\dfrac{\partial u_1}{\partial r}$ in eq. (4.1-1) is programmed.

```
#
# u1rr, Neumann BC, r=0,ru
   u1rr=matrix(0,nrow=nr,ncol=nz);
   for(iz in 1:nz){
   for(ir in 1:nr){
     if(ir==1){
       u1rr[1,iz]=4*(u1[2,iz]-u1[1,iz])/dr2;}
     if(ir==nr){
       u1rr[nr,iz]=2*(u1[nr-1,iz]-u1[nr,iz])/dr2;}
     if((ir>1)&(ir<nr)){
       u1rr[ir,iz]=(u1[ir+1,iz]-2*u1[ir,iz]+u1[ir-1,iz])/dr2+
```

```
                    (1/r[ir])*(u1[ir+1,iz]-u1[ir-1,iz])/(2*dr);}
    }
}
```

This coding requires some additional explanation.

- The radial derivative term $\dfrac{1}{r}\dfrac{\partial u_1}{\partial r}$ in eq. (4.1-1) is indeterminate at $r = 0$ with the application of BC (4.1-3).

$$\frac{1}{r}\frac{\partial u_1}{\partial r}\Big|_{r=0} = \frac{0}{0}$$

- Application of l'Hospital's rule to this indeterminate term gives

$$\frac{1}{r}\frac{\partial u_1}{\partial r}\Big|_{r=0} = \frac{\partial^2 u_1}{\partial r^2}\Big|_{r=0}$$

and the radial group in eq. (4.1-1) is

$$\frac{\partial^2 u_1}{\partial r^2} + \frac{1}{r}\frac{\partial u_1}{\partial r}\Big|_{r=0}$$

$$= 2\frac{\partial^2 u_1}{\partial r^2}$$

- The FD approximation of the second derivative at $r = 0$ is

$$\frac{\partial^2 u_1}{\partial r^2}\Big|_{r=0}$$

$$\approx \frac{u_1(r = \Delta r, z, t) - 2u(r = 0, z, t) + u_1(r = -\Delta r, z, t)}{\Delta r^2}$$

- The fictitious term $u_1(r = -\Delta r, z, t)$ is approximated from BC (4.1-3) with a two point FD centered at $r = 0$

$$\frac{u_1(r = \Delta r, z, t) - u_1(r = -\Delta r, z, t)}{2\Delta r} = 0$$

or

$$u_1(r = -\Delta r, z, t) \approx u_1(r = \Delta r, z, t)$$

- The second derivative is then approximated as

$$\frac{\partial^2 u_1}{\partial r^2}\Big|_{r=0}$$

$$\approx 2\frac{u_1(r = \Delta r, z, t) - u(r = 0, z, t)}{\Delta r^2}$$

- The radial group in eq. (4.1-1) at $r = 0$ is therefore

$$\frac{\partial^2 u_1}{\partial r^2} + \frac{1}{r}\frac{\partial u_1}{\partial r}\Big|_{r=0}$$

$$\approx 4\frac{u_1(r=\Delta r, z, t) - u_1(r=0, z, t)}{\Delta r^2}$$

which is programmed as

```
if(ir==1){
  u1rr[1,iz]=4*(u1[2,iz]-u1[1,iz])/dr2;}
```

• The radial derivative group $\frac{\partial^2 u_1}{\partial r^2} + \frac{1}{r}\frac{\partial u_1}{\partial r}$ in eq. (4.1-1) at $r = r_u = 1$ with application of homogeneous Neumann BC (4.1-4) is

$$\frac{\partial^2 u_1}{\partial r^2} + \frac{1}{r}\frac{\partial u_1}{\partial r}\Big|_{r=r_u=1} = \frac{\partial^2 u_1}{\partial r^2}$$

As before (at $r = 0$), this derivative is approximated as

$$\frac{\partial^2 u_1}{\partial r^2}\Big|_{r=r_u=1} \approx 2\frac{u_1(r=r_u - \Delta r, z, t) - u(r=r_u, z, t)}{\Delta r^2}$$

and is programmed as

```
if(ir==nr){
  u1rr[nr,iz]=2*(u1[nr-1,iz]-u1[nr,iz])/dr2;}
```

which includes BC (4.1-4).

• For the interior points $r > r_l$, $r < r_u$, the radial group in eq. (4.1-1) is approximated with FDs as

$$\frac{\partial^2 u_1}{\partial r^2} + \frac{1}{r}\frac{\partial u_1}{\partial r}$$

$$\approx \frac{u_1(r+\Delta r, z, t) - 2u_1(r, z, t) + u_1(r-\Delta r, z, t)}{\Delta r^2}$$

$$+\frac{1}{r}\frac{u_1(r+\Delta r, z, t) - u_1(r-\Delta r, z, t)}{2\Delta r}$$

and is programmed as

```
if((ir>1)&(ir<nr)){
  u1rr[ir,iz]=(u1[ir+1,iz]-2*u1[ir,iz]+u1[ir-1,iz])/dr2+
              (1/r[ir])*(u1[ir+1,iz]-u1[ir-1,iz])/(2*dr);}
```

This completes the programming of the radial group in eq. (4.1-1) over the 2D $n_r \times n_z$ grid in r, z. This group can now be used in the MOL programming of eq. (4.1-1) as indicated subsequently.

• The axial derivative $\frac{\partial^2 u_1}{\partial z^2}$ in eq. (4.1-1) is programmed.

```
#
# u1zz, Neumann BC, z=0,zu
  u1zz=matrix(0,nrow=nr,ncol=nz);
  for(iz in 1:nz){
  for(ir in 1:nr){
    if(iz==1){
      u1zz[ir,1]=2*(u1[ir,2]-u1[ir,1])/dz2;}
    if(iz==nz){
      u1zz[ir,nz]=2*(u1[ir,nz-1]-u1[ir,nz])/dz2;}
    if((iz>1)&(iz<nz)){
      u1zz[ir,iz]=(u1[ir,iz+1]-2*u1[ir,iz]+u1[ir,iz-1])/dz2;}
  }
  }
```

This programming follows analogously from the previous coding in r with homogeneous Neumann BCs at $z = z_l$, $z = z_u$

z=zl

```
  if(iz==1){
    u1zz[ir,1]=2*(u1[ir,2]-u1[ir,1])/dz2;}
```

z=zu

```
  if(iz==nz){
    u1zz[ir,nz]=2*(u1[ir,nz-1]-u1[ir,nz])/dz2;}
```

- The MOL programming of eq. (4.1-1) gives the derivative $\dfrac{\partial u_1(r,z,t)}{\partial t}$, including the Michaelis-Menten source term q_1.
- An analogous group of FD approximations of the radial and axial derivatives in eq. (4.2-1) is next.

```
#
# u2rr, Neumann BC, r=0,ru
  u2rr=matrix(0,nrow=nr,ncol=nz);
  for(iz in 1:nz){
  for(ir in 1:nr){
    if(ir==1){
      u2rr[1,iz]=4*(u2[2,iz]-u2[1,iz])/dr2;}
    if(ir==nr){
      u2rr[nr,iz]=2*(u2[nr-1,iz]-u2[nr,iz])/dr2;}
    if((ir>1)&(ir<nr)){
      u2rr[ir,iz]=(u2[ir+1,iz]-2*u2[ir,iz]+u2[ir-1,iz])/dr2+
                  (1/r[ir])*(u2[ir+1,iz]-u2[ir-1,iz])/(2*dr);}
```

```
    }
    }
#
# u2zz, Neumann BC, z=0,zu
  u2zz=matrix(0,nrow=nr,ncol=nz);
  for(iz in 1:nz){
  for(ir in 1:nr){
    if(iz==1){
      u2zz[ir,1]=2*(u2[ir,2]-u2[ir,1])/dz2;}
    if(iz==nz){
      u2zz[ir,nz]=2*(u2[ir,nz-1]-u2[ir,nz])/dz2;}
    if((iz>1)&(iz<nz)){
      u2zz[ir,iz]=(u2[ir,iz+1]-2*u2[ir,iz]+u2[ir,iz-1])/dz2;}
  }
  }
```

- At this point, all of the RHS terms of eqs. (4.1-1), (4.2-1) are available numerically so that the MOL programming of these two simultaneous PDEs can be completed.

```
  }
#
# u1t, u2t
  u1t=matrix(0,nrow=nr,ncol=nz);
  u2t=matrix(0,nrow=nr,ncol=nz);
  for(iz in 1:nz){
  for(ir in 1:nr){
#
#   Michaelis-Menten
    q1=pQm1*u1[ir,iz]/(cm1+u1[ir,iz])*u2[ir,iz];
    q2=pQm2*u2[ir,iz]/(cm2+u2[ir,iz])*u1[ir,iz];
#
#   PDE
    u1t[ir,iz]=D1r*u1rr[ir,iz]+D1z*u1zz[ir,iz]-q1;
    u2t[ir,iz]=D2r*u2rr[ir,iz]+D2z*u2zz[ir,iz]-q2;
  }
  }
```

Note in particular the programming of q_1, q_2 of eqs. (4.1-7), (4.2-7).

- The 462 ODE derivatives are placed in the vector ut for return to lsodes to take the next step in t along the solution.

```
#
# 2D matrices to 1D vector
  ut=rep(0,2*nr*nz)
```

```
for(iz in 1:nz){
for(ir in 1:nr){
  ut[(iz-1)*nr+ir]        =u1t[ir,iz];
  ut[(iz-1)*nr+ir+nr*nz]=u2t[ir,iz];
}
}
```

- The counter for the calls to pde2a is incremented and returned to the main program of Listing 4.1 by <<-.

```
#
# Increment calls to pde2a
  ncall <<- ncall+1;
```

- The vector ut is returned as a list as required by lsodes. c is the R vector utility.

```
#
# Return derivative vector
  return(list(c(ut)));
  }
```

The final } concludes pde2a.

This completes the discussion of pde2a. The output from the main program of Listing 4.1 and ODE/MOL routine pde2a of Listing 4.2 is considered next.

4.1.3 Numerical, graphical output

Abbreviated output for ncase=1 (set in Listing 4.1) follows.

Table 4.1 Numerical output from Listings 4.1, 4.2, ncase=1.

[1] 11			
[1] 463			
t	z	(iz-1)nr+ir	u1(r=0,z,t)
0.0	0.000	1.0	1.000e+00
0.0	0.050	12.0	1.000e+00
0.0	0.100	23.0	1.000e+00
0.0	0.150	34.0	1.000e+00
.	.		.
.	.		.
.	.		.

continued on next page

Table 4.1   (continued)

```
          Output for z = 0.2 to 0.8 removed
                    .                    .
                    .                    .
                    .                    .
     0.0    0.850      188.0          1.000e+00
     0.0    0.900      199.0          1.000e+00
     0.0    0.950      210.0          1.000e+00
     0.0    1.000      221.0          1.000e+00


      t       z      (iz-1)nr+ir   u2(r=0,z,t)
     0.0    0.000        1.0         1.000e+00
     0.0    0.050       12.0         1.000e+00
     0.0    0.100       23.0         1.000e+00
     0.0    0.150       34.0         1.000e+00
                    .                    .
                    .                    .
                    .                    .
          Output for z = 0.2 to 0.8 removed
                    .                    .
                    .                    .
                    .                    .
     0.0    0.850      188.0          1.000e+00
     0.0    0.900      199.0          1.000e+00
     0.0    0.950      210.0          1.000e+00
     0.0    1.000      221.0          1.000e+00


      t       z      (iz-1)nr+ir   u1(r=0,z,t)
     0.1    0.000        1.0         9.518e-01
     0.1    0.050       12.0         9.518e-01
     0.1    0.100       23.0         9.518e-01
     0.1    0.150       34.0         9.518e-01
                    .                    .
                    .                    .
                    .                    .
          Output for z = 0.2 to 0.8 removed
                    .                    .
                    .                    .
                    .                    .
     0.1    0.850      188.0          9.518e-01
     0.1    0.900      199.0          9.518e-01
     0.1    0.950      210.0          9.518e-01
     0.1    1.000      221.0          9.518e-01


      t       z      (iz-1)nr+ir   u2(r=0,z,t)
     0.1    0.000        1.0         9.518e-01
     0.1    0.050       12.0         9.518e-01
```

continued on next page

Table 4.1 (continued)

t	z	(iz-1)nr+ir	u1(r=0,z,t)
0.1	0.100	23.0	9.518e-01
0.1	0.150	34.0	9.518e-01
.	.		.
.	.		.
.	.		.

Output for z = 0.2 to 0.8 removed

.	.		.
.	.		.
.	.		.
0.1	0.850	188.0	9.518e-01
0.1	0.900	199.0	9.518e-01
0.1	0.950	210.0	9.518e-01
0.1	1.000	221.0	9.518e-01
.	.		.
.	.		.
.	.		.

Output for t = 0.2 to 0.9 removed

.	.		.
.	.		.
.	.		.

t	z	(iz-1)nr+ir	u1(r=0,z,t)
1.0	0.000	1.0	6.422e-01
1.0	0.050	12.0	6.422e-01
1.0	0.100	23.0	6.422e-01
1.0	0.150	34.0	6.422e-01
.	.		.
.	.		.
.	.		.

Output for z = 0.2 to 0.8 removed

.	.		.
.	.		.
.	.		.
1.0	0.850	188.0	6.422e-01
1.0	0.900	199.0	6.422e-01
1.0	0.950	210.0	6.422e-01
1.0	1.000	221.0	6.422e-01

t	z	(iz-1)nr+ir	u2(r=0,z,t)
1.0	0.000	1.0	6.422e-01
1.0	0.050	12.0	6.422e-01
1.0	0.100	23.0	6.422e-01
1.0	0.150	34.0	6.422e-01
.	.		.
.	.		.
.	.		.

continued on next page

Table 4.1 (continued)

Output for z = 0.2 to 0.8 removed			
	.		.
	.		.
	.		.
1.0	0.850	188.0	6.422e-01
1.0	0.900	199.0	6.422e-01
1.0	0.950	210.0	6.422e-01
1.0	1.000	221.0	6.422e-01
ncall =	500		

We can note the following details about this output.

- 11 t output points as the first dimension of the solution matrix out from lsodes as programmed in the main program of Listing 4.1 (with nout=11).
- The solution matrix out returned by lsodes has 463 elements as a second dimension. The first element is the value of t. Elements 2 to 463 are $u_1(r,z,t)$, $u_2(r,z,t)$ from eqs. (4.1-1), (4.2-1) (for the spatial grid with $n_r = 11$, $n_z = 21$, $2(n_r)(n_z) = 2(11)(21) = 462$ points).
- The solutions are displayed for t=0,1/10=0.1,...,1 as programmed in Listing 4.1.
- ICs (4.1-2), (4.2-2) are confirmed ($t = 0$).
- The solutions $u_1(r = 0,z,t)$, $u_2(r = 0,z,t)$ are the same since eqs. (4.1), (4.2), including the numerical parameters defined in Listing 4.1, are identical. This is an important special case since a difference in the two solutions would indicate a programming error.
- The solutions $u_1(r = 0,z,t)$, $u_2(r = 0,z,t)$ do not vary with z since the ICs $u_1(r,z,t = 0) = 1$, $u_2(r,z,t = 0) = 1$ and BCs (4.1-3,4,5,6), (4.2-3,4,5,6) are consistent (with BCs (4.1-5), (4.2-5) replaced with homogeneous Neumann BCs as programmed in pde2a in Listing 4.2). This is an important check since a variation of the solutions with z would indicate a programming error.
- The decay in the solutions with t results from the source terms $q_1 > 0$, $q_2 > 0$ in eqs. (4.1-1), (4.2-1). $q_1 = 0$, $q_2 = 0$ in Listings 4.1, 4.2 would result in a constant solution that does not change from the ICs $u_1(r = 0,z,t = 0) = 1$, $u_2(r = 0,z,t = 0) = 1$. This is also an important special case that is left as an exercise.
- The computational effort as indicated by ncall = 500 is modest so that lsodes computed the solution to eqs. (4.1), (4.2) efficiently.

The graphical output is in Figs. 4.1.
The solution in Fig. 4.1-1 does not vary in z and decays from $q_1 > 0$.

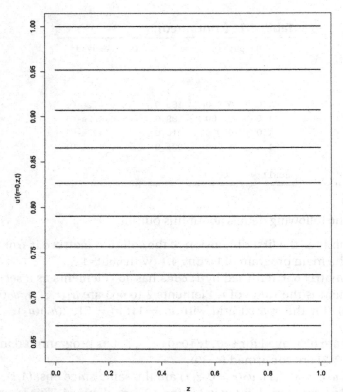

FIGURE 4.1-1 $u_1(r=0, z, t)$ from eqs. (4.1), (4.2), 2D, ncase=1.

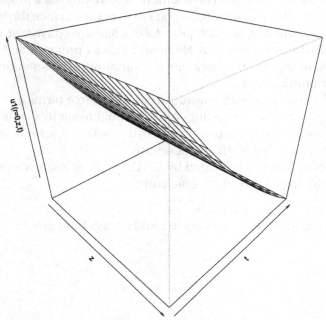

FIGURE 4.1-2 $u_1(r=0, z, t)$ from eqs. (4.1), (4.2), 3D, ncase=1.

Since the solutions for $u_1(r = 0, z, t)$, $u_2(r = 0, z, t)$ are identical, the plotted numerical solution for $u_2(r = 0, z, t)$ is not included here.

For ncase=2 the numerical output follows.

Table 4.2   Numerical output from Listings 4.1, 4.2, ncase=2.

[1] 11

[1] 463

t	z	(iz-1)nr+ir	u1(r=0,z,t)
0.0	0.000	1.0	0.000e+00
0.0	0.050	12.0	0.000e+00
0.0	0.100	23.0	0.000e+00
0.0	0.150	34.0	0.000e+00
.	.	.	.
.	.	.	.
.	.	.	.

Output for z = 0.2 to 0.8 removed

t	z	(iz-1)nr+ir	u1(r=0,z,t)
.	.	.	.
.	.	.	.
.	.	.	.
0.0	0.850	188.0	0.000e+00
0.0	0.900	199.0	0.000e+00
0.0	0.950	210.0	0.000e+00
0.0	1.000	221.0	0.000e+00

t	z	(iz-1)nr+ir	u2(r=0,z,t)
0.0	0.000	1.0	0.000e+00
0.0	0.050	12.0	0.000e+00
0.0	0.100	23.0	0.000e+00
0.0	0.150	34.0	0.000e+00
.	.	.	.
.	.	.	.
.	.	.	.

Output for z = 0.2 to 0.8 removed

t	z	(iz-1)nr+ir	u2(r=0,z,t)
.	.	.	.
.	.	.	.
.	.	.	.
0.0	0.850	188.0	0.000e+00
0.0	0.900	199.0	0.000e+00
0.0	0.950	210.0	0.000e+00
0.0	1.000	221.0	0.000e+00

t	z	(iz-1)nr+ir	u1(r=0,z,t)
0.1	0.000	1.0	1.553e-01
0.1	0.050	12.0	1.639e-01

continued on next page

Table 4.2 (continued)

```
0.1    0.100    23.0     1.883e-01
0.1    0.150    34.0     2.240e-01
          .                  .
          .                  .
          .                  .
    Output for z = 0.2 to 0.8 removed
          .                  .
          .                  .
          .                  .
0.1    0.850   188.0     2.240e-01
0.1    0.900   199.0     1.883e-01
0.1    0.950   210.0     1.639e-01
0.1    1.000   221.0     1.553e-01

 t      z    (iz-1)nr+ir  u2(r=0,z,t)
0.1    0.000     1.0     1.553e-01
0.1    0.050    12.0     1.639e-01
0.1    0.100    23.0     1.883e-01
0.1    0.150    34.0     2.240e-01
          .                  .
          .                  .
          .                  .
    Output for z = 0.2 to 0.8 removed
          .                  .
          .                  .
          .                  .
0.1    0.850   188.0     2.240e-01
0.1    0.900   199.0     1.883e-01
0.1    0.950   210.0     1.639e-01
0.1    1.000   221.0     1.553e-01
          .                  .
          .                  .
          .                  .
    Output for t = 0.2 to 0.9 removed
          .                  .
          .                  .
          .                  .
 t      z    (iz-1)nr+ir  u1(r=0,z,t)
1.0    0.000     1.0     3.394e-01
1.0    0.050    12.0     3.396e-01
1.0    0.100    23.0     3.402e-01
1.0    0.150    34.0     3.410e-01
          .                  .
          .                  .
          .                  .
    Output for z = 0.2 to 0.8 removed
```
continued on next page

Table 4.2   (continued)

.	.		.
.	.		.
.	.		.
1.0	0.850	188.0	3.410e-01
1.0	0.900	199.0	3.402e-01
1.0	0.950	210.0	3.396e-01
1.0	1.000	221.0	3.394e-01
t	z	(iz-1)nr+ir	u2(r=0,z,t)
1.0	0.000	1.0	3.394e-01
1.0	0.050	12.0	3.396e-01
1.0	0.100	23.0	3.402e-01
1.0	0.150	34.0	3.410e-01
	.		.
	.		.
	.		.

Output for t = 0.2 to 0.9 removed

	.		.
	.		.
	.		.
1.0	0.850	188.0	3.410e-01
1.0	0.900	199.0	3.402e-01
1.0	0.950	210.0	3.396e-01
1.0	1.000	221.0	3.394e-01

ncall =   582

We can note the following details about this output.

- ICs (4.1-2), (4.2-2) are confirmed, $u_1(r = 0, z, t = 0) = 0$, $u_2(r = 0, z, t = 0) = 0$.
- The solutions $u_1(r = 0, z, t)$, $u_2(r = 0, z, t)$ are the same since eqs. (4.1), (4.2), including the numerical parameters defined in Listing 4.1, are identical. This is an important special case since a difference in the two solutions would indicate a programming error.
- The solution remains symmetric around $z = 0.5$. This is an important test since any asymmetry would indicate a programming error.
- The homogeneous Neumann BCs are verified and the solution approaches a steady state value of approximately 0.34.
- The computational effort as indicated by `ncall` = 582 is modest so that `lsodes` computed the solution to eqs. (4.1), (4.2) efficiently.

The graphical output is in Figs. 4.2.

The solutions display the properties discussed for Table 4.2 (symmetry around $z = 0.5$ and approach to a steady state solution).

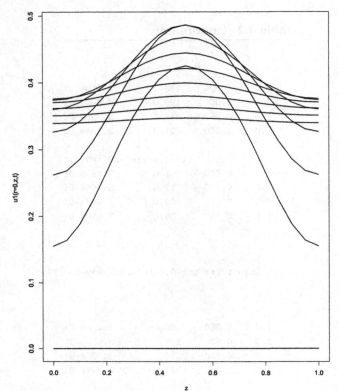

FIGURE 4.2-1 $u_1(r = 0, z, t)$ from eqs. (4.2), (4.3), 2D, ncase=2.

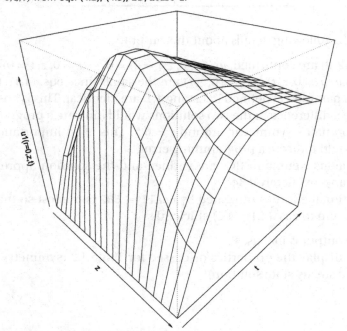

FIGURE 4.2-2 $u_1(r = 0, z, t)$ from eqs. (4.1), (4.2), 3D, ncase=2.

Since the solutions for $u_1(r = 0, z, t)$, $u_2(r = 0, z, t)$ are identical, the plotted numerical solution for $u_2(r = 0, z, t)$ is not included here.

This concludes the discussion of Listings 4.1, 4.2 for which homogeneous Neumann BCs are imposed at $r = r_l, r_u, z = z_l, z_u$ (to test various solution properties with `ncase=1,2` in the coding as discussed previously). The case with a Dirichlet BC at $z = z_l$, as specified by $g_1(r, t)$, $g_2(r, t)$ in eqs. (4.1-5), (4.2-5) is now considered.

4.1.4 Main program, application

The main program follows.

```
#
# 2D TC model
#
# Delete previous workspaces
  rm(list=ls(all=TRUE))
#
# Access ODE integrator
  library("deSolve");
#
# Access functions for numerical solution
  setwd("f:/tissue engineering/chap4");
  source("pde2b.R");
#
# Parameters
#
# Component 1
  D1r=0.1;
  D1z=0.1;
  p1=1;
  Qm1=1;
  cm1=1;
  pQm1=p1*Qm1;
  u1b=1;
  u10=0;
#
# Component 2
  D2r=0.1;
  D2z=0.1;
  p2=1;
  Qm2=1;
  cm2=1;
```

```
  pQm2=p2*Qm2;
  u2b=1;
  u20=0;
#
# Spatial grid (in r)
  nr=11;rl=0;ru=1;dr=(ru-rl)/(nr-1);dr2=dr^2;
  r=seq(from=rl,to=ru,by=dr);
#
# Spatial grid (in z)
  nz=21;zl=0;zu=1;dz=(zu-zl)/(nz-1);dz2=dz^2;
  z=seq(from=zl,to=zu,by=dz);
#
# Independent variable for ODE integration
  t0=0;tf=1;nout=11;
  tout=seq(from=t0,to=tf,by=(tf-t0)/(nout-1));
#
# Initial conditions (t=0)
  u0=rep(0,2*nr*nz);
  for(iz in 1:nz){
  for(ir in 1:nr){
    u0[(iz-1)*nr+ir]       =u10;
    u0[(iz-1)*nr+ir+nr*nz]=u20;
  }
  }
   ncall=0;
#
# ODE integration
  out=lsodes(y=u0,times=tout,func=pde2b,
      sparsetype ="sparseint",rtol=1e-6,
      atol=1e-6,maxord=5);
  nrow(out)
  ncol(out)
#
# Arrays for plotting numerical solution
  u1=matrix(0,nrow=nz,ncol=nout);
  u2=matrix(0,nrow=nz,ncol=nout);
  for(it in 1:nout){
    cat(sprintf("\n      t      z      (iz-1)nr+ir  u1(r=0,z,t)\n"));
    ir=1;
    for(iz in 1:nz){
      u1[iz,it]=out[it,(iz-1)*nr+ir+1];
      u1[1,it]=u1b;
```

```
          cat(sprintf("%7.1f%9.3f%9.1f%16.3e\n",
              tout[it],z[iz],(iz-1)*nr+ir,u1[iz,it]));
      }
      cat(sprintf("\n      t        z      (iz-1)nr+ir  u2(r=0,z,t)\n"));
      for(iz in 1:nz){
        u2[iz,it]=out[it,(iz-1)*nr+ir+1+nr*nz];
        u2[1,it]=u2b;
        cat(sprintf("%7.1f%9.3f%9.1f%16.3e\n",
            tout[it],z[iz],(iz-1)*nr+ir,u2[iz,it]));
      }
  }
#
# Calls to ODE routine
  cat(sprintf("\n\n ncall = %5d\n\n",ncall));
#
# Plot PDE solutions
#
# u1(z,t)
# 2D
  par(mfrow=c(1,1));
  matplot(x=z[2:nz],y=u1[2:nz,],type="l",xlab="z",
    ylab="u1(r=0,z,t)",lty=1,main="",lwd=2,col="black");
#
# 3D
  persp(z,tout,u1,theta=45,phi=30,xlim=c(zl,zu),
    ylim=c(t0,tf),xlab="z",ylab="t",zlab="u1(r=0,z,t)");
#
# u2(z,t)
# 2D
  par(mfrow=c(1,1));
  matplot(x=z[2:nz],y=u2[2:nz,],type="l",xlab="z",
    ylab="u2(r=0,z,t)",lty=1,main="",lwd=2,col="black");
#
# 3D
  persp(z,tout,u2,theta=45,phi=30,xlim=c(zl,zu),
    ylim=c(t0,tf),xlab="z",ylab="t",zlab="u2(r=0,z,t)");
```

Listing 4.3: Main program for eqs. (4.1), (4.2), Dirichlet BC.

We can note the following details about Listing 4.3 (with some repetition of the discussion of Listing 3.3 so that the following explanation is self contained).

- Previous workspaces are deleted.

```
#
# 2D TC model
#
# Delete previous workspaces
  rm(list=ls(all=TRUE))
```

- The R ODE integrator library deSolve is accessed [1].

```
#
# Access ODE integrator
  library("deSolve");
#
# Access functions for numerical solution
  setwd("f:/tissue engineering/chap4");
  source("pde2b.R");
```

Then the directory with the files for the solution of eqs. (4.1), (4.2) is designated. Note that setwd (set working directory) uses / rather than the usual \. The ordinary differential equation/method of lines (ODE/MOL) routine, pde2b, is accessed through setwd.

- The model parameters are specified numerically.

```
#
# Parameters
#
# Component 1
  D1r=0.1;
  D1z=0.1;
  p1=1;
  Qm1=1;
  cm1=1;
  pQm1=p1*Qm1;
  u1b=1;
  u10=0;
#
# Component 2
  D2r=0.1;
  D2z=0.1;
  p2=1;
  Qm2=1;
  cm2=1;
  pQm2=p2*Qm2;
  u2b=1;
  u20=0;
```

In particular,

- u10,u20: Initial values $u_1(r,z,t=0)=u_{10}=0$, $u_2(r,z,t=0)=u_{20}=0$ in ICs (4.1-2), (4.2-2).
- A spatial grid in r for eqs. (4.1-1) (4.2-1) is defined with 11 points so that r = $0,1/10=0.1,\ldots,1$. The stem cell length is a normalized value, $r=r_u=1$.

```
#
# Spatial grid (in r)
  nr=11;rl=0;ru=1;dr=(ru-rl)/(nr-1);dr2=dr^2;
  r=seq(from=rl,to=ru,by=dr);
```

- A spatial grid in z for eqs. (4.1-1) (4.2-1) is defined with 21 points so that z = 0,1/20= $0.05,\ldots,1$. The stem cell length is a normalized value, $z=z_u=1$.

```
#
# Spatial grid (in z)
  nz=21;zl=0;zu=1;dz=(zu-zl)/(nz-1);dz2=dz^2;
  z=seq(from=zl,to=zu,by=dz);
```

- An interval in t is defined for 11 output points, so that tout=0,1/10=0.1,...,1. The time scale is normalized with $t_f=1$ specified as the final time that is considered appropriate, e.g., minutes, days.

```
# Independent variable for ODE integration
  t0=0;tf=1;nout=11;
  tout=seq(from=t0,to=tf,by=(tf-t0)/(nout-1));
```

- ICs (4.1-2), (4.2-2) are implemented.

```
#
# Initial conditions (t=0)
  u0=rep(0,2*nr*nz);
  for(iz in 1:nz){
  for(ir in 1:nr){
    u0[(iz-1)*nr+ir]        =u10;
    u0[(iz-1)*nr+ir+nr*nz]=u20;
  }
  }
  ncall=0;
```

Also, the counter for the calls to pde2b is initialized.
- The system of 2*nr*nz=2*11*21=462 ODEs is integrated by the library integrator lsodes (available in deSolve, [1]). As expected, the inputs to lsodes are the ODE function,

pde2b, the IC vector u0, and the vector of output values of t, tout. The length of u0 (462) informs lsodes how many ODEs are to be integrated. func,y,times are reserved names.

```
#
# ODE integration
  out=lsodes(y=u0,times=tout,func=pde2b,
      sparsetype ="sparseint",rtol=1e-6,
      atol=1e-6,maxord=5);
  nrow(out)
  ncol(out)
```

nrow,ncol confirm the dimensions of out.

- $u_1(r = 0, z, t)$, $u_2(r = 0, z, t)$ are placed in matrices for plotting.

```
#
# Arrays for plotting numerical solution
  u1=matrix(0,nrow=nz,ncol=nout);
  u2=matrix(0,nrow=nz,ncol=nout);
  for(it in 1:nout){
    cat(sprintf("\n      t        z      (iz-1)nr+ir  u1(r=0,z,t)\n"));
    ir=1;
    for(iz in 1:nz){
      u1[iz,it]=out[it,(iz-1)*nr+ir+1];
      u1[1,it]=u1b;
      cat(sprintf("%7.1f%9.3f%9.1f%16.3e\n",
          tout[it],z[iz],(iz-1)*nr+ir,u1[iz,it]));
     }
    cat(sprintf("\n      t        z      (iz-1)nr+ir  u2(r=0,z,t)\n"));
    for(iz in 1:nz){
      u2[iz,it]=out[it,(iz-1)*nr+ir+1+nr*nz];
      u2[1,it]=u2b;
      cat(sprintf("%7.1f%9.3f%9.1f%16.3e\n",
          tout[it],z[iz],(iz-1)*nr+ir,u2[iz,it]));
     }
   }
```

The offset +1 is required since the first element of the solution vectors in out is the value of t and the 2 to 463 elements are the 462 values of u_1, u_2. These dimensions from the preceding calls to nrow,ncol are confirmed in the subsequent output.

The boundary values $u_1(r = 0, z = 0, t)$, $u_2(r = 0, z = 0, t)$ are set by algebraic equations in pde2b, but lsodes returns only solutions to ODEs. So the boundary values are set in the main program above as u1[1,it]=u1b, u2[1,it]=u2b.

- The number of calls to pde2b is displayed at the end of the solution.

```
#
# Calls to ODE routine
  cat(sprintf("\n\n ncall = %5d\n\n",ncall));
```

- $u_1(r = 0, z, t)$, $u_2(r = 0, z, t)$ are plotted in 2D against z and parametrically in t with the R utility matplot, and in 3D with the R utility persp. par(mfrow=c(1,1)) specifies a 1×1 matrix of plots, that is, one plot on a page.

```
#
# Calls to ODE routine
  cat(sprintf("\n\n ncall = %5d\n\n",ncall));
#
# Plot PDE solutions
#
# u1(z,t)
# 2D
  par(mfrow=c(1,1));
  matplot(x=z[2:nz],y=u1[2:nz,],type="l",xlab="z",
    ylab="u1(r=0,z,t)",lty=1,main="",lwd=2,col="black");
#
# 3D
  persp(z,tout,u1,theta=45,phi=30,xlim=c(zl,zu),
    ylim=c(t0,tf),xlab="z",ylab="t",zlab="u1(r=0,z,t)");
#
# u2(z,t)
# 2D
  par(mfrow=c(1,1));
  matplot(x=z[2:nz],y=u2[2:nz,],type="l",xlab="z",
    ylab="u2(r=0,z,t)",lty=1,main="",lwd=2,col="black");
#
# 3D
  persp(z,tout,u2,theta=45,phi=30,xlim=c(zl,zu),
    ylim=c(t0,tf),xlab="z",ylab="t",zlab="u2(r=0,z,t)");
```

By using [2:nz], the discontinuity between the IC $u_1(r = 0, z, t = 0) = 0$ and the BC $u_1(r = 0, z = 0, t) = 1$ is not included in the 2D plot (Fig. 4.3-1). Including the discontinuity is left as an exercise.

This completes the discussion of the main program for eqs. (4.1), (4.2). The ODE/MOL routine pde2b called by lsodes from the main program of Listing 4.3 for the numerical MOL integration of eqs. (4.1), (4.2) is next.

4.1.5 ODE/MOL routine

The ODE/MOL routine, pde2b, called by lsodes from the main program of Listing 4.3 follows.

```
  pde2b=function(t,u,parm){
#
# Function pde2b computes the t derivative
# of u1(r,z,t), u2(r,z,t)
#
# 1D vector to 2D matrices
  u1=matrix(0,nrow=nr,ncol=nz);
  u2=matrix(0,nrow=nr,ncol=nz);
  for(iz in 1:nz){
  for(ir in 1:nr){
    u1[ir,iz]=u[(iz-1)*nr+ir];
    u2[ir,iz]=u[(iz-1)*nr+ir+nr*nz];
  }
  }
#
# u1rr, Neumann BC, r=0,ru
  u1rr=matrix(0,nrow=nr,ncol=nz);
  for(iz in 1:nz){
  for(ir in 1:nr){
    if(ir==1){
      u1rr[1,iz]=4*(u1[2,iz]-u1[1,iz])/dr2;}
    if(ir==nr){
      u1rr[nr,iz]=2*(u1[nr-1,iz]-u1[nr,iz])/dr2;}
    if((ir>1)&(ir<nr)){
      u1rr[ir,iz]=(u1[ir+1,iz]-2*u1[ir,iz]+u1[ir-1,iz])/dr2+
                  (1/r[ir])*(u1[ir+1,iz]-u1[ir-1,iz])/(2*dr);}
  }
  }
#
# u1zz, Dirichlet, z=zl, Neumann BC, z=zu,
  u1zz=matrix(0,nrow=nr,ncol=nz);
  for(iz in 1:nz){
  for(ir in 1:nr){
    if(iz==1){
      u1[ir,1]=u1b};
    if(iz==nz){
      u1zz[ir,nz]=2*(u1[ir,nz-1]-u1[ir,nz])/dz2;}
```

```
      if((iz>1)&(iz<nz)){
        u1zz[ir,iz]=(u1[ir,iz+1]-2*u1[ir,iz]+u1[ir,iz-1])/dz2;}
    }
    }
#
# u2rr, Neumann BC, r=0,ru
  u2rr=matrix(0,nrow=nr,ncol=nz);
  for(iz in 1:nz){
  for(ir in 1:nr){
    if(ir==1){
      u2rr[1,iz]=4*(u2[2,iz]-u2[1,iz])/dr2;}
    if(ir==nr){
      u2rr[nr,iz]=2*(u2[nr-1,iz]-u2[nr,iz])/dr2;}
    if((ir>1)&(ir<nr)){
      u2rr[ir,iz]=(u2[ir+1,iz]-2*u2[ir,iz]+u2[ir-1,iz])/dr2+
                  (1/r[ir])*(u2[ir+1,iz]-u2[ir-1,iz])/(2*dr);}
  }
  }
#
# u1zz, Dirichlet, z=zl, Neumann BC, z=zu,
  u2zz=matrix(0,nrow=nr,ncol=nz);
  for(iz in 1:nz){
  for(ir in 1:nr){
    if(iz==1){
      u2[ir,1]=u2b};
    if(iz==nz){
      u2zz[ir,nz]=2*(u2[ir,nz-1]-u2[ir,nz])/dz2;}
    if((iz>1)&(iz<nz)){
      u2zz[ir,iz]=(u2[ir,iz+1]-2*u2[ir,iz]+u2[ir,iz-1])/dz2;}
  }
  }
#
# u1t, u2t
  u1t=matrix(0,nrow=nr,ncol=nz);
  u2t=matrix(0,nrow=nr,ncol=nz);
  for(iz in 1:nz){
  for(ir in 1:nr){
#
#   Michaelis-Menten
    q1=pQm1*u1[ir,iz]/(cm1+u1[ir,iz])*u2[ir,iz];
    q2=pQm2*u2[ir,iz]/(cm2+u2[ir,iz])*u1[ir,iz];
#
```

```
#    PDE
    u1t[ir,iz]=D1r*u1rr[ir,iz]+D1z*u1zz[ir,iz]-q1;
    u2t[ir,iz]=D2r*u2rr[ir,iz]+D2z*u2zz[ir,iz]-q2;
    u1t[ir,1]=0;
    u2t[ir,1]=0;
  }
  }
#
# 2D matrices to 1D vector
  ut=rep(0,2*nr*nz)
  for(iz in 1:nz){
  for(ir in 1:nr){
    ut[(iz-1)*nr+ir]        =u1t[ir,iz];
    ut[(iz-1)*nr+ir+nr*nz]=u2t[ir,iz];
  }
  }
#
# Increment calls to pde2b
  ncall <<- ncall+1;
#
# Return derivative vector
  return(list(c(ut)));
  }
```

Listing 4.4: ODE/MOL routine for eqs. (4.1), (4.2), Dirichlet BC.

We can note the following details about Listing 4.4 (with some repetition of the discussion of Listing 4.2 so that the following explanation is self contained).

• The function is defined.

```
    pde2b=function(t,u,parm){
#
# Function pde2b computes the t derivative
# of u1(r,z,t), u2(r,z,t)
```

t is the current value of t in eqs. (4.1), (4.2). u is the 462-vector of ODE/PDE dependent variables. parm is an argument to pass parameters to pde2b (unused, but required in the argument list). The arguments must be listed in the order stated to properly interface with lsodes called in the main program of Listing 4.3. The derivative vector of the LHS of eqs. (4.1-1), (4.2-1) is calculated and returned to lsodes as explained subsequently.

• The input vector u is placed in matrices u1, u2 to facilitate the programming of eqs. (4.1), (4.2).

```
#
# 1D vector to 2D matrices
  u1=matrix(0,nrow=nr,ncol=nz);
  u2=matrix(0,nrow=nr,ncol=nz);
  for(iz in 1:nz){
  for(ir in 1:nr){
    u1[ir,iz]=u[(iz-1)*nr+ir];
    u2[ir,iz]=u[(iz-1)*nr+ir+nr*nz];
  }
  }
```

- The radial group in eq. (4.1-1), $\dfrac{\partial^2 u_1}{\partial^2 r} + \dfrac{1}{r}\dfrac{\partial u_1}{\partial r}$, is programmed (a detailed explanation of this code follows after Listing 3.2).

```
#
# u1rr, Neumann BC, r=0,ru
  u1rr=matrix(0,nrow=nr,ncol=nz);
  for(iz in 1:nz){
  for(ir in 1:nr){
    if(ir==1){
      u1rr[1,iz]=4*(u1[2,iz]-u1[1,iz])/dr2;}
    if(ir==nr){
      u1rr[nr,iz]=2*(u1[nr-1,iz]-u1[nr,iz])/dr2;}
    if((ir>1)&(ir<nr)){
      u1rr[ir,iz]=(u1[ir+1,iz]-2*u1[ir,iz]+u1[ir-1,iz])/dr2+
                  (1/r[ir])*(u1[ir+1,iz]-u1[ir-1,iz])/(2*dr);}
  }
  }
```

- The derivative $\dfrac{\partial^2 u_1}{\partial z^2}$, in eq. (4.1-1) is programmed.

```
#
# u1zz, Dirichlet, z=zl, Neumann BC, z=zu,
  u1zz=matrix(0,nrow=nr,ncol=nz);
  for(iz in 1:nz){
  for(ir in 1:nr){
    if(iz==1){
      u1[ir,1]=u1b};
    if(iz==nz){
      u1zz[ir,nz]=2*(u1[ir,nz-1]-u1[ir,nz])/dz2;}
    if((iz>1)&(iz<nz)){
      u1zz[ir,iz]=(u1[ir,iz+1]-2*u1[ir,iz]+u1[ir,iz-1])/dz2;}
```

```
        }
        }
```

This coding requires some additional explanation.

- The derivative $\dfrac{\partial^2 u_1}{\partial z^2}$ in eq. (4.1-1) is approximated with a three point centered (in z) finite difference (FD).

$$\frac{\partial^2 u_1(r,z,t)}{\partial z^2} \approx \frac{(u_1(r,z+\Delta z,t) - 2u_1(r,z,t) + u_1(r,z-\Delta z,t))}{\Delta z^2} + O(\Delta z^2) \qquad (4.3\text{-}1)$$

$O(\Delta z^2)$ indicates that the error in the FD approximation is second order in Δz. The variation in the numerical solution of eq. (4.1-1) can be studied as a function of the FD increment Δz by varying nz in Listing 4.3.
- Dirichlet BC (4.1-5) is applied at $z = z_l = 0$.

$$u_1(r, z = 0, t) = u_{1b} \qquad (4.3\text{-}2)$$

and is programmed as

```
        if(iz==1){
          u1[ir,1]=u1b;}
```

- At $z = z_u$, Neumann BC (4.1-6) is used to eliminate the fictitious value $u_1(r, z_u + \Delta z, t)$, and eq. (4.3-1) is

$$\frac{\partial^2 u_1(z_u,t)}{\partial z^2} \approx \frac{2(u_1(z_u - \Delta z, t) - u_1(z_u, t))}{\Delta z^2} \qquad (4.3\text{-}3)$$

that is programmed as

```
        if(iz==nz){
          u1zz[ir,nz]=2*(u1[ir,nz-1]-u1[ir,nz])/dz2;}
```

- For the interior points $z_l < z < z_u$, eq. (4.1-1) is programmed as

```
        if((iz>1)&(iz<nz)){
          u1zz[ir,iz]=(u1[ir,iz+1]-2*u1[ir,iz]+u1[ir,iz-1])/dz2;}
```

$\dfrac{\partial^2 u_1}{\partial z^2}$ is now available for use in the programming of eq. (4.1-1).
- Analogous programming of the RHS derivatives of eq. (4.2-1) is next.

```
#
# u2rr, Neumann BC, r=0,ru
  u2rr=matrix(0,nrow=nr,ncol=nz);
  for(iz in 1:nz){
  for(ir in 1:nr){
    if(ir==1){
```

```
            u2rr[1,iz]=4*(u2[2,iz]-u2[1,iz])/dr2;}
        if(ir==nr){
            u2rr[nr,iz]=2*(u2[nr-1,iz]-u2[nr,iz])/dr2;}
        if((ir>1)&(ir<nr)){
            u2rr[ir,iz]=(u2[ir+1,iz]-2*u2[ir,iz]+u2[ir-1,iz])/dr2+
                        (1/r[ir])*(u2[ir+1,iz]-u2[ir-1,iz])/(2*dr);}
    }
    }
#
# u1zz, Dirichlet, z=zl, Neumann BC, z=zu,
    u2zz=matrix(0,nrow=nr,ncol=nz);
    for(iz in 1:nz){
    for(ir in 1:nr){
        if(iz==1){
            u2[ir,1]=u2b};
        if(iz==nz){
            u2zz[ir,nz]=2*(u2[ir,nz-1]-u2[ir,nz])/dz2;}
        if((iz>1)&(iz<nz)){
            u2zz[ir,iz]=(u2[ir,iz+1]-2*u2[ir,iz]+u2[ir,iz-1])/dz2;}
    }
    }
```

• Eqs. (4.1-1), (4.2-1) are programmed in the MOL format.

```
#
# u1t, u2t
    u1t=matrix(0,nrow=nr,ncol=nz);
    u2t=matrix(0,nrow=nr,ncol=nz);
    for(iz in 1:nz){
    for(ir in 1:nr){
#
#   Michaelis-Menten
    q1=pQm1*u1[ir,iz]/(cm1+u1[ir,iz])*u2[ir,iz];
    q2=pQm2*u2[ir,iz]/(cm2+u2[ir,iz])*u1[ir,iz];
#
#   PDE
    u1t[ir,iz]=D1r*u1rr[ir,iz]+D1z*u1zz[ir,iz]-q1;
    u2t[ir,iz]=D2r*u2rr[ir,iz]+D2z*u2zz[ir,iz]-q2;
    u1t[ir,1]=0;
    u2t[ir,1]=0;
    }
    }
```

- The 462 ODE derivatives are placed in the vector ut for return to lsodes to take the next step in *t* along the solution.

```
#
# 2D matrices to 1D vector
  ut=rep(0,2*nr*nz)
  for(iz in 1:nz){
  for(ir in 1:nr){
    ut[(iz-1)*nr+ir]       =u1t[ir,iz];
    ut[(iz-1)*nr+ir+nr*nz]=u2t[ir,iz];
  }
  }
```

- The counter for the calls to pde2b is incremented and returned to the main program of Listing 3.3 by <<-.

```
#
# Increment calls to pde2b
  ncall <<- ncall+1;
```

- The vector ut is returned as a list as required by lsodes. c is the R vector utility.

```
#
# Return derivative vector
  return(list(c(ut)));
  }
```

The final } concludes pde1b.

This completes the discussion of pde2b. The output from the main program of Listing 4.3 and ODE/MOL routine pde2b of Listing 4.4 is considered next.

4.1.6 Numerical, graphical output

Table 4.3   Numerical output from Listings 4.3, 4.4, Dirichlet BC.

```
[1] 11

[1] 463
```

t	z	(iz-1)nr+ir	u1(r=0,z,t)
0.0	0.000	1.0	1.000e+00
0.0	0.050	12.0	0.000e+00
0.0	0.100	23.0	0.000e+00

continued on next page

Table 4.3 (continued)

t	z	(iz-1)nr+ir	u(r=0,z,t)
0.0	0.150	34.0	0.000e+00
	.		.
	.		.
	.		.

Output for z = 0.2 to 0.8 removed

	.		.
	.		.
	.		.
0.0	0.850	188.0	0.000e+00
0.0	0.900	199.0	0.000e+00
0.0	0.950	210.0	0.000e+00
0.0	1.000	221.0	0.000e+00

t	z	(iz-1)nr+ir	u2(r=0,z,t)
0.0	0.000	1.0	1.000e+00
0.0	0.050	12.0	0.000e+00
0.0	0.100	23.0	0.000e+00
0.0	0.150	34.0	0.000e+00
	.		.
	.		.
	.		.

Output for z = 0.2 to 0.8 removed

	.		.
	.		.
	.		.
0.0	0.850	188.0	0.000e+00
0.0	0.900	199.0	0.000e+00
0.0	0.950	210.0	0.000e+00
0.0	1.000	221.0	0.000e+00

t	z	(iz-1)nr+ir	u1(r=0,z,t)
0.1	0.000	1.0	1.000e+00
0.1	0.050	12.0	7.164e-01
0.1	0.100	23.0	4.726e-01
0.1	0.150	34.0	2.855e-01
	.		.
	.		.
	.		.

Output for z = 0.2 to 0.8 removed

	.		.
	.		.
	.		.
0.1	0.850	188.0	5.919e-08
0.1	0.900	199.0	1.238e-08
0.1	0.950	210.0	2.550e-09
0.1	1.000	221.0	9.398e-10

continued on next page

Table 4.3 (continued)

t	z	(iz-1)nr+ir	u2(r=0,z,t)
0.1	0.000	1.0	1.000e+00
0.1	0.050	12.0	7.164e-01
0.1	0.100	23.0	4.726e-01
0.1	0.150	34.0	2.855e-01
.	.		.
.	.		.
.	.		.

Output for z = 0.2 to 0.8 removed

.	.		.
.	.		.
.	.		.
0.1	0.850	188.0	5.919e-08
0.1	0.900	199.0	1.238e-08
0.1	0.950	210.0	2.550e-09
0.1	1.000	221.0	9.398e-10
.	.		.
.	.		.
.	.		.

Output for t = 0.2 to 0.9 removed

.	.		.
.	.		.
.	.		.

t	z	(iz-1)nr+ir	u1(r=0,z,t)
1.0	0.000	1.0	1.000e+00
1.0	0.050	12.0	8.831e-01
1.0	0.100	23.0	7.773e-01
1.0	0.150	34.0	6.815e-01
.	.		.
.	.		.
.	.		.

Output for z = 0.2 to 0.8 removed

.	.		.
.	.		.
1.0	0.850	188.0	6.112e-02
1.0	0.900	199.0	5.296e-02
1.0	0.950	210.0	4.814e-02
1.0	1.000	221.0	4.654e-02

t	z	(iz-1)nr+ir	u2(r=0,z,t)
1.0	0.000	1.0	1.000e+00
1.0	0.050	12.0	8.831e-01
1.0	0.100	23.0	7.773e-01
1.0	0.150	34.0	6.815e-01

continued on next page

Table 4.3 (continued)

.			.
.			.
.			.

Output for z = 0.2 to 0.8 removed

.			.
.			.
.			.
1.0	0.850	188.0	6.112e-02
1.0	0.900	199.0	5.296e-02
1.0	0.950	210.0	4.814e-02
1.0	1.000	221.0	4.654e-02

ncall =   603

We can note the following details about this output.

- 11 t output points as the first dimension of the solution matrix out from lsodes as programmed in the main program of Listing 4.1 (with nout=11).
- The solution matrix out returned by lsodes has 463 elements as a second dimension. The first element is the value of t. Elements 2 to 463 are $u_1(r, z, t)$, $u_2(r, z, t)$ from eqs. (4.1), (4.2) (for the spatial grid with $n_r = 11$, $n_z = 21$, $2(n_r)(n_z) = 2(11)(21) = 462$ points).
- Since the programming for $u_2(r, z, t)$ is the same as for $u_1(z, z, t)$, including the parameters in Listing 4.3, the two solutions are the same. This is an important special case since a difference in the two PDE variables would indicate a programming error.
- The solution is displayed for t=0,1/10=0.1,...,1 as programmed in Listing 3.1.
- The solutions $u_1(r = 0, z, t)$, $u_2(r = 0, z, t)$ are displayed for (iz-1)*nr+ir+1 with iz=1,...,21, ir=1 programmed in Listing 3.1.
- ICs (4.1-2), (4.2-2) are confirmed (at $t = 0$).
  Note in particular that BCs (4.1-5), (4.2-5) are verified (e.g., $u_1(r = 0, z = 0, t) = g_1(r = 0, t) = 1$).
- The computational effort as indicated by ncall = 603 is modest so that lsodes computed the solution to eqs. (4.1), (4.2) efficiently.

The graphical output is in Figs. 4.3.

This solution displays the properties discussed for Table 4.3 (BC (4.1-5) that moves the solution from IC (4.1-2)).

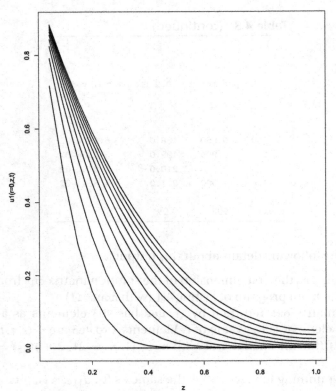

FIGURE 4.3-1 $u_1(r = 0, z, t)$ from eqs. (4.1), 2D, Dirichlet BC.

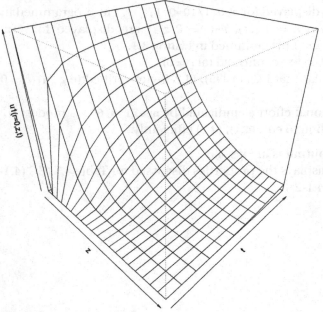

FIGURE 4.3-2 $u_1(r = 0, z, t)$ from eqs. (4.1), 3D, Dirichlet BC.

The graphical output for $u_2(r, z, t)$ is not included here since it is the same as for $u_1(r, z, t)$ (Figs. 4.3-1,2).

This concludes the discussion of Listings 4.3, 4.4 for which Dirichlet BCs are specified at $z = z_l$ and Neumann BCs at $z = z_u$.

As a concluding application, the programming in Listings 4.3, 4.4 is used to study the effect of a diffusion limited nutrient. Specifically, the diffusivities D2r, D2z specified in Listing 4.3 are changed from

```
D2r=0.1;
D2z=0.1;
```

to

```
D2r=0.01;
D2z=0.01;
```

Table 4.4 Numerical output from Listings 4.3, 4.4, reduced D2r,D2z.

[1] 11

[1] 463

t	z	(iz-1)nr+ir	u1(r=0,z,t)
0.0	0.000	1.0	1.000e+00
0.0	0.050	12.0	0.000e+00
0.0	0.100	23.0	0.000e+00
0.0	0.150	34.0	0.000e+00
.	.	.	.
.	.	.	.
.	.	.	.

Output for z = 0.2 to 0.8 removed

.	.	.	.
.	.	.	.
0.0	0.850	188.0	0.000e+00
0.0	0.900	199.0	0.000e+00
0.0	0.950	210.0	0.000e+00
0.0	1.000	221.0	0.000e+00

t	z	(iz-1)nr+ir	u2(r=0,z,t)
0.0	0.000	1.0	1.000e+00
0.0	0.050	12.0	0.000e+00
0.0	0.100	23.0	0.000e+00
0.0	0.150	34.0	0.000e+00
.	.	.	.
.	.	.	.
.	.	.	.

Output for z = 0.2 to 0.8 removed

continued on next page

Table 4.4   (continued)

0.0	0.850	188.0	0.000e+00
0.0	0.900	199.0	0.000e+00
0.0	0.950	210.0	0.000e+00
0.0	1.000	221.0	0.000e+00

t	z	(iz-1)nr+ir	u1(r=0,z,t)
0.1	0.000	1.0	1.000e+00
0.1	0.050	12.0	7.207e-01
0.1	0.100	23.0	4.773e-01
0.1	0.150	34.0	2.888e-01

Output for z = 0.2 to 0.8 removed

t	z	(iz-1)nr+ir	u1(r=0,z,t)
0.1	0.850	188.0	5.932e-08
0.1	0.900	199.0	1.241e-08
0.1	0.950	210.0	2.558e-09
0.1	1.000	221.0	9.436e-10

t	z	(iz-1)nr+ir	u2(r=0,z,t)
0.1	0.000	1.0	1.000e+00
0.1	0.050	12.0	2.749e-01
0.1	0.100	23.0	4.761e-02
0.1	0.150	34.0	5.911e-03

Output for z = 0.2 to 0.8 removed

t	z	(iz-1)nr+ir	u2(r=0,z,t)
0.1	0.850	188.0	2.436e-22
0.1	0.900	199.0	5.536e-24
0.1	0.950	210.0	1.200e-25
0.1	1.000	221.0	4.977e-27

Output for t = 0.2 to 0.9 removed

continued on next page

Table 4.4 (continued)

t	z	(iz-1)nr+ir	u1(r=0,z,t)
1.0	0.000	1.0	1.000e+00
1.0	0.050	12.0	8.991e-01
1.0	0.100	23.0	8.067e-01
1.0	0.150	34.0	7.205e-01
.	.	.	.
.	.	.	.
.	.	.	.

Output for z = 0.2 to 0.8 removed

.	.	.	.
.	.	.	.
.	.	.	.
1.0	0.850	188.0	6.733e-02
1.0	0.900	199.0	5.810e-02
1.0	0.950	210.0	5.264e-02
1.0	1.000	221.0	5.084e-02

t	z	(iz-1)nr+ir	u2(r=0,z,t)
1.0	0.000	1.0	1.000e+00
1.0	0.050	12.0	6.387e-01
1.0	0.100	23.0	3.841e-01
1.0	0.150	34.0	2.164e-01
.	.	.	.
.	.	.	.
.	.	.	.

Output for z = 0.2 to 0.8 removed

.	.	.	.
.	.	.	.
.	.	.	.
1.0	0.850	188.0	4.852e-08
1.0	0.900	199.0	1.028e-08
1.0	0.950	210.0	2.145e-09
1.0	1.000	221.0	7.980e-10

ncall =   599

We can note the following details about this output.

- The number of output points is again nout=11 with 2*nr*nz+1=2*11*21+1=463 values in the solution array out returned by lsodes in the main program of Listing 4.3.
- BCs (4.1-5), (4.2-5) are confirmed, e.g., at $t = 1$,

t	z	(iz-1)nr+ir	u1(r=0,z,t)
1.0	0.000	1.0	1.000e+00

t	z	(iz-1)nr+ir	u2(r=0,z,t)

```
        1.0      0.000        1.0          1.000e+00
```

- $u_2(r, z, t) << u_1(r, z, t)$ with the reduction in D2r, D2z. Specifically, $u_2(r = 0, z \to z_u = 1, t)$ has low values, e.g., from Table 4.4,

```
    t          z      (iz-1)nr+ir   u1(r=0,z,t)
    1.0      1.000      221.0        5.084e-02


    t          z      (iz-1)nr+ir   u2(r=0,z,t)
    1.0      1.000      221.0        7.980e-10
```

In other words, the diffusion limited nutrient has a pronounced effect (on the solution $u_2(r, z, t)$).

The graphical output is in Figs. 4.4.

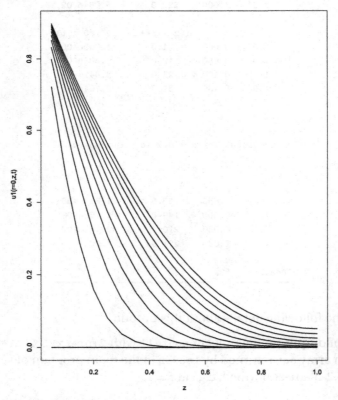

FIGURE 4.4-1 $u_1(r = 0, z, t)$ from eqs. (4.1), 2D, reduced D2r, D2z.

This solution displays the properties discussed for Table 4.4 (BC (4.1-5) that moves the solution from IC (4.1-2)).

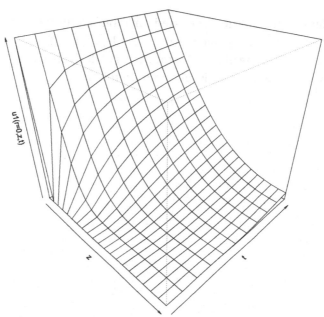

FIGURE 4.4-2 $u_1(r = 0, z, t)$ from eqs. (4.1), 3D, reduced D2r, D2z.

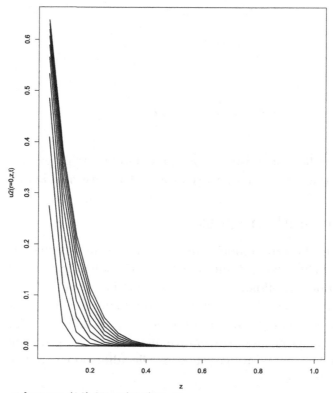

FIGURE 4.4-3 $u_2(r = 0, z, t)$ from eqs. (4.2), 2D, reduced D2r, D2z.

This solution displays the properties discussed for Table 4.4 (BC (4.2-5) that moves the solution from IC (4.2-2)). $u_2(r, z, t)$ is below $u_1(r, z, t)$ of Fig. 4.1-1 (from the diffusion limited nutrient). The apparent gridding effect (discontinuous z derivative) can be studied for spatial convergence by using nz>21. This is left as an exercise.

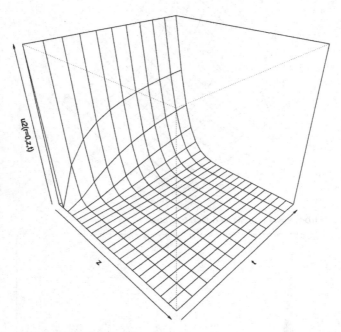

FIGURE 4.4-4 $u_2(r = 0, z, t)$ from eqs. (4.2), 3D, reduced D2r,D2z.

This example application illustrates experimentation with parameter values. Variations and extensions of the model can also be considered as illustrated in the next chapter.

4.2 Summary and conclusions

2D O_2 and nutrient diffusion models are implemented in this chapter for: (1) two Neumann BCs, (2) one Dirichlet and one Neumann BC applied to the spatial domain of the stem cells seeded on a scaffold. The R routines are based on the MOL with the spatial derivatives approximated by finite differences. For (2), the increase of O_2 and nutrient concentrations provides essential conditions for the stem cell differentiation that could be the basis for tissue engineering and regenerative medicine. As an application of the two PDE model, the reduced nutrient through rate controlled diffusion is studied through the variation of the nutrient diffusivity.

In the next chapter, additional PDEs are added to the two PDE model for differentiation of the stem cells that are energized by O_2 and a nutrient (metabolism).

Reference

[1] K. Soetaert, J. Cash, F. Mazzia, Solving Differential Equations in R, Springer-Verlag, Heidelberg, Germany, 2012.

5

Stem cell metabolism and differentiation

Introduction

In this concluding chapter, the PDE model pertains to the stem cell metabolism and differentiation, formulated and implemented as two sets of PDEs.

5.1 Metabolism

The PDE dependent variables $u_{1m}(r, z, t)$, $u_{2m}(r, z, t)$ are $u_1(r, z, t)$, $u_2(r, z, t)$ of eqs. (4.1), (4.2) with m added to designate metabolism. The independent variables r, z, t are defined in Table 5.1.

Table 5.1 Dependent, independent variables of PDEs (5.1).

normalized variable	physical interpretation
u_{1m}	O_2 level in seeded stem cells
u_{2m}	nutrient level in seeded stem cells
r	radial distance in stem cells; lower boundary, $r = r_l = 0$; upper boundary at external surface, $r = r_u$
z	axial distance in stem cells; lower boundary, $z = z_l$; upper boundary at scaffold surface, $z = z_u$
t	time

$$\frac{\partial u_{1m}}{\partial t} = D_{1m,r}\left(\frac{\partial^2 u_{1m}}{\partial r^2} + \frac{1}{r}\frac{\partial u_{1m}}{\partial r}\right) + D_{1m,z}\frac{\partial^2 u_{1m}}{\partial z^2} - q_1 \tag{5.1-1}$$

$$\frac{\partial u_{2m}}{\partial t} = D_{2m,r}\left(\frac{\partial^2 u_{2m}}{\partial r^2} + \frac{1}{r}\frac{\partial u_{2m}}{\partial r}\right) + D_{2m,z}\frac{\partial^2 u_{2m}}{\partial z^2} - q_2 \tag{5.1-2}$$

$q_1 = q_1(u_{1m}(r, z, t), u_{2m}(r, t))$ is a two component Michaelis-Menten rate.

$$q_1 = pQm_1\frac{u_{1m}}{cm_1 + u_{1m}}u_{2m} \qquad (5.1\text{-}3)$$

$q_2 = q_2(u_{1m}(r, z, t), u_{2m}(r, t))$ is a two component Michaelis-Menten rate.

$$q_2 = pQm_2\frac{u_{2m}}{cm_2 + u_{2m}}u_{1m} \qquad (5.1\text{-}4)$$

The auxiliary conditions (IC plus BCs) for u_{1m} (eq. (5.1-1)) follow.

$$u_{1m}(r, z, t = 0) = f_{1m}(r, z) \qquad (5.2\text{-}1)$$

$$\frac{\partial u_{1m}(r = 0, z, t)}{\partial r} = 0 \qquad (5.2\text{-}2)$$

$$\frac{\partial u_{1m}(r = r_u, z, t)}{\partial r} = 0 \qquad (5.2\text{-}3)$$

$$u_{1m}(r, z = 0, t) = g_{1m}(r, t) \qquad (5.2\text{-}4)$$

$$\frac{\partial u_{1m}(r, z = z_u, t)}{\partial z} = 0 \qquad (5.2\text{-}5)$$

$f_{1m}(r, z)$, $g_{1m}(r, t)$ are functions to be specified.

The auxiliary conditions (IC plus BCs) for u_{2m} (eq. (5.1-2)) follow.

$$u_{2m}(r, z, t = 0) = f_{2m}(r, z) \qquad (5.3\text{-}1)$$

$$\frac{\partial u_{2m}(r = 0, z, t)}{\partial r} = 0 \qquad (5.3\text{-}2)$$

$$\frac{\partial u_{2m}(r = r_u, z, t)}{\partial r} = 0 \qquad (5.3\text{-}3)$$

$$u_{2m}(r, z = 0, t) = g_{2m}(r, t) \qquad (5.3\text{-}4)$$

$$\frac{\partial u_{2m}(r, z = z_u, t)}{\partial z} = 0 \qquad (5.3\text{-}5)$$

$f_{2m}(r, z)$, $g_{2m}(r, t)$ are functions to be specified.

q_1, q_2 couple the PDEs, eqs. (5.1-1), (5.1-2), since these are functions of both $u_{1m}(r, z, t)$ and $u_{2m}(r, z, t)$. In other words, eqs. (5.1-1) and (5.1-2) must be integrated (solved) simultaneously and not sequentially (one at a time).

5.2 Differentiation

The PDE dependent variables in addition to u_{1m}, u_{2m} are u_1, u_2, u_3, u_4, u_5, u_6 [1] defined in Table 5.2. The independent variables r, z, t are defined in Table 5.1.

Table 5.2   Dependent variables of PDEs (5.4).

normalized variable	physical interpretation
u_1	stem cell density
u_2	transit-amplifying (TA) cell density
u_3	terminally differentiated (TD) cell density
u_4	signaling (regulatory) biomolecule 1 concentration produced by u_1, u_2, u_3, u_6
u_5	signaling (regulatory) biomolecule 2 concentration produced by u_1, u_2, u_3, u_6
u_6	signaling (regulatory) biomolecule 3 concentration produced by u_4, u_5
p_0, p_1	regulated proliferation probabilities
p_{0m}, p_{1m}	maximal replication probabilities
γ_4, γ_5	u_4, u_5 sensitivity coefficients, respectively
m, n	Hill coefficients

The PDEs for $u_1(r, z, t)$ to $u_6(r, z, t)$ follow.

$$\frac{\partial u_1}{\partial t} = D_{1,r}\left(\frac{\partial^2 u_1}{\partial r^2} + \frac{1}{r}\frac{\partial u_1}{\partial r}\right) + D_{1,z}\frac{\partial^2 u_1}{\partial z^2}$$
$$+a_{11}(2p_0 - 1)u_1 + q_1 \tag{5.4-1}$$

$$\frac{\partial u_2}{\partial t} = D_{2,r}\left(\frac{\partial^2 u_2}{\partial r^2} + \frac{1}{r}\frac{\partial u_2}{\partial r}\right) + D_{2,z}\frac{\partial^2 u_2}{\partial z^2}$$
$$+a_{21}[2(1 - p_0)u_1] + a_{22}[(2p_1 - 1)u_2] \tag{5.4-2}$$

$$\frac{\partial u_3}{\partial t} = D_{3,r}\left(\frac{\partial^2 u_3}{\partial r^2} + \frac{1}{r}\frac{\partial u_3}{\partial r}\right) + D_{3,z}\frac{\partial^2 u_3}{\partial z^2}$$
$$+a_{31}[2(1 - p_1)u_2] - a_{32}u_3 \tag{5.4-3}$$

where

$$p_0 = \frac{p_{0m}}{1 + (\gamma_4 u_4(z,t))^m} \tag{5.4-a}$$

$$p_1 = \frac{p_{1m}}{1 + (\gamma_5 u_5(z,t))^n} \tag{5.4-b}$$

$$\frac{\partial u_4}{\partial t} = D_{4,r}\left(\frac{\partial^2 u_4}{\partial r^2} + \frac{1}{r}\frac{\partial u_4}{\partial r}\right) + D_{4,z}\frac{\partial^2 u_4}{\partial z^2}$$

$$+a_{41}u_1 + a_{42}u_2 + a_{43}u_3 - a_{44}u_4 - a_{45}u_4u_6 \tag{5.4-4}$$

$$\frac{\partial u_5}{\partial t} = D_{5,r}\left(\frac{\partial^2 u_5}{\partial r^2} + \frac{1}{r}\frac{\partial u_5}{\partial r}\right) + D_{5,z}\frac{\partial^2 u_5}{\partial z^2}$$

$$+a_{51}u_1 + a_{52}u_2 + a_{53}u_3 - a_{54}u_5 - a_{55}u_5u_6 \tag{5.4-5}$$

$$\frac{\partial u_6}{\partial t} = D_{6,r}\left(\frac{\partial^2 u_6}{\partial r^2} + \frac{1}{r}\frac{\partial u_6}{\partial r}\right) D_{6,z}\frac{\partial^2 u_6}{\partial z^2}$$

$$-a_{61}u_6 - a_{62}u_4u_6 - a_{63}u_5u_6 \tag{5.4-6}$$

The auxiliary conditions (IC plus BCs) for u_1 (for eq. (5.4-1)) follow.

$$u_1(r, z, t=0) = f_1(r, z) \tag{5.5-1}$$

$$\frac{\partial u_1(r=r_l=0, z, t)}{\partial r} = 0 \tag{5.5-2}$$

$$\frac{\partial u_1(r=r_u, z, t)}{\partial r} = 0 \tag{5.5-3}$$

$$u_1(r, z=z_l, t) = g_1(r, t) \tag{5.5-4}$$

$$\frac{\partial u_1(r, z=z_u, t)}{\partial r} = 0 \tag{5.5-5}$$

The auxiliary conditions (IC plus BCs) for u_2 (for eq. (5.4-2)) follow.

$$u_2(r, z, t=0) = f_2(r, z) \tag{5.6-1}$$

$$\frac{\partial u_2(r=r_l=0, z, t)}{\partial r} = 0 \tag{5.6-2}$$

$$\frac{\partial u_2(r=r_u, z, t)}{\partial r} = 0 \tag{5.6-3}$$

$$u_2(r, z=z_l, t) = g_2(r, t) \tag{5.6-4}$$

$$\frac{\partial u_2(r, z=z_u, t)}{\partial r} = 0 \tag{5.6-5}$$

The auxiliary conditions (IC plus BCs) for u_3 (for eq. (5.4-3)) follow.

$$u_3(r, z, t=0) = f_3(r, z) \tag{5.7-1}$$

$$\frac{\partial u_3(r=r_l=0, z, t)}{\partial r} = 0 \tag{5.7-2}$$

$$\frac{\partial u_3(r=r_u, z, t)}{\partial r} = 0 \tag{5.7-3}$$

$$u_3(r, z=z_l, t) = g_3(r, t) \tag{5.7-4}$$

$$\frac{\partial u_3(r, z=z_u, t)}{\partial r} = 0 \tag{5.7-5}$$

The auxiliary conditions (IC plus BCs) for u_4 (for eq. (5.4-4)) follow.

$$u_4(r, z, t = 0) = f_4(r, z) \tag{5.8-1}$$

$$\frac{\partial u_4(r = r_l = 0, z, t)}{\partial r} = 0 \tag{5.8-2}$$

$$\frac{\partial u_4(r = r_u, z, t)}{\partial r} = 0 \tag{5.8-3}$$

$$u_4(r, z = z_l, t) = g_4(r, t) \tag{5.8-4}$$

$$\frac{\partial u_4(r, z = z_u, t)}{\partial r} = 0 \tag{5.8-5}$$

The auxiliary conditions (IC plus BCs) for u_5 (for eq. (5.4-5)) follow.

$$u_5(r, z, t = 0) = f_5(r, z) \tag{5.9-1}$$

$$\frac{\partial u_5(r = r_l = 0, z, t)}{\partial r} = 0 \tag{5.9-2}$$

$$\frac{\partial u_5(r = r_u, z, t)}{\partial r} = 0 \tag{5.9-3}$$

$$u_5(r, z = z_l, t) = g_5(r, t) \tag{5.9-4}$$

$$\frac{\partial u_5(r, z = z_u, t)}{\partial r} = 0 \tag{5.9-5}$$

The auxiliary conditions (IC plus BCs) for u_6 (for eq. (5.4-6)) follow.

$$u_6(r, z, t = 0) = f_6(r, z) \tag{5.10-1}$$

$$\frac{\partial u_6(r = r_l = 0, z, t)}{\partial r} = 0 \tag{5.10-2}$$

$$\frac{\partial u_6(r = r_u, z, t)}{\partial r} = 0 \tag{5.10-3}$$

$$u_6(r, z = z_l, t) = g_6(r, t) \tag{5.10-4}$$

$$\frac{\partial u_6(r, z = z_u, t)}{\partial r} = 0 \tag{5.10-5}$$

5.3 Implementation of metabolism and differentiation model

Eqs. (5.1) to (5.10) constitute the eight PDE model for stem cell metabolism and differentiation implemented in the following R routines.

5.3.1 Main program

The main program for eqs. (5.1), (5.4) follows.

```
#
# 2D TC model
#
# Delete previous workspaces
  rm(list=ls(all=TRUE))
#
# Access ODE integrator
  library("deSolve");
#
# Access functions for numerical solution
  setwd("f:/tissue engineering/chap5");
  source("pde2b.R");
#
# Parameters
#
# u1m
  D1mr=0.5;
  D1mz=0.5;
  p1=1;
  Qm1=1;
  cm1=1;
  pQm1=p1*Qm1;
  u1mb=1;
  u1m0=0;
#
# u2m
  D2mr=0.5;
  D2mz=0.5;
  p2=1;
  Qm2=1;
  cm2=1;
  pQm2=p2*Qm2;
  u2mb=1;
  u2m0=0;
#
# u1
  D1r=0.25;
  D1z=0.25;
  a11=1;
  u1b=1;
  u10=0;
#
```

```
# u2
  D2r=0.25;
  D2z=0.25;
  a21=1;a22=1;
  u2b=0;
  u20=0;
#
# u3
  D3r=0.25;
  D3z=0.25;
  a31=1;a32=1;
  u3b=0;
  u30=0;
#
# u4
  D4r=0.25;
  D4z=0.25;
  a41=1;a42=1;
  a43=1;a44=1;
  a45=1;
  u4b=0;
  u40=0;
#
# u5
  D5r=0.25;
  D5z=0.25;
  a51=1;a52=1;
  a53=1;a54=1;
  a55=1;
  u5b=0;
  u50=0;
#
# u6
  D6r=0.25;
  D6z=0.25;
  a61=1;a62=1;
  a63=1;
  u6b=0;
  u60=0;
#
# Parameters probability functions
  p0m=1;p1m=0.1;
```

```
   g4=1;  g5=1;
    m=2;   n=2;
#
# Spatial grid (in r)
  nr=6;rl=0;ru=1;dr=(ru-rl)/(nr-1);dr2=dr^2;
  r=seq(from=rl,to=ru,by=dr);
#
# Spatial grid (in z)
  nz=21;zl=0;zu=1;dz=(zu-zl)/(nz-1);dz2=dz^2;
  z=seq(from=zl,to=zu,by=dz);
#
# Independent variable for ODE integration
  t0=0;tf=1;nout=11;
  tout=seq(from=t0,to=tf,by=(tf-t0)/(nout-1));
#
# Initial conditions (t=0)
  u0=rep(0,8*nr*nz);
  for(iz in 1:nz){
  for(ir in 1:nr){
    u0[(iz-1)*nr+ir]          =u1m0;
    u0[(iz-1)*nr+ir+nr*nz]    =u2m0;
    u0[(iz-1)*nr+ir+2*nr*nz]= u10;
    u0[(iz-1)*nr+ir+3*nr*nz]= u20;
    u0[(iz-1)*nr+ir+4*nr*nz]= u30;
    u0[(iz-1)*nr+ir+5*nr*nz]= u40;
    u0[(iz-1)*nr+ir+6*nr*nz]= u50;
    u0[(iz-1)*nr+ir+7*nr*nz]= u60;
  }
  }
   ncall=0;
#
# ODE integration
  out=lsodes(y=u0,times=tout,func=pde2b,
      sparsetype ="sparseint",rtol=1e-6,
      atol=1e-6,maxord=5);
  nrow(out)
  ncol(out)
#
# Arrays for displaying, plotting numerical
# solutions
  u1m=matrix(0,nrow=nz,ncol=nout);
  u2m=matrix(0,nrow=nz,ncol=nout);
```

```
    u1=matrix(0,nrow=nz,ncol=nout);
    u2=matrix(0,nrow=nz,ncol=nout);
    u3=matrix(0,nrow=nz,ncol=nout);
    u4=matrix(0,nrow=nz,ncol=nout);
    u5=matrix(0,nrow=nz,ncol=nout);
    u6=matrix(0,nrow=nz,ncol=nout);
#
# Save solutions from lsodes
  for(it in 1:nout){
  for(iz in 1:nz){
      u1m[iz,it]=out[it,(iz-1)*nr+ir+1];
      u1m[1,it]=u1mb;
      u2m[iz,it]=out[it,(iz-1)*nr+ir+1+ nr*nz];
      u2m[1,it]=u2mb;
      u1[iz,it]=out[it,(iz-1)*nr+ir+1+2*nr*nz];
      u1[1,it]=u1b;
      u2[iz,it]=out[it,(iz-1)*nr+ir+1+3*nr*nz];
      u2[1,it]=u2b;
      u3[iz,it]=out[it,(iz-1)*nr+ir+1+4*nr*nz];
      u3[1,it]=u3b;
      u4[iz,it]=out[it,(iz-1)*nr+ir+1+5*nr*nz];
      u4[1,it]=u4b;
      u5[iz,it]=out[it,(iz-1)*nr+ir+1+6*nr*nz];
      u5[1,it]=u5b;
      u6[iz,it]=out[it,(iz-1)*nr+ir+1+7*nr*nz];
      u6[1,it]=u6b;
  }
  }
#
# Display solutions
  iv=seq(from=1,to=nout,by=5);
  for(it in iv){
  ivz=seq(from=1,to=nz,by=5);
    ir=1;
    cat(sprintf("\n      t       z        u1m(r=0,z,t)\n"));
    for(iz in ivz){
      cat(sprintf("%7.1f%9.3f%16.3e\n",
          tout[it],z[iz],u1m[iz,it]));
    }
    cat(sprintf("\n      t       z        u2m(r=0,z,t)\n"));
    for(iz in ivz){
      cat(sprintf("%7.1f%9.3f%16.3e\n",
```

```
                tout[it],z[iz],u2m[iz,it]));
    }
    cat(sprintf("\n       t        z           u1(r=0,z,t)\n"));
    for(iz in ivz){
      cat(sprintf("%7.1f%9.3f%16.3e\n",
          tout[it],z[iz],u1[iz,it]));
    }
    cat(sprintf("\n       t        z           u2(r=0,z,t)\n"));
    for(iz in ivz){
      cat(sprintf("%7.1f%9.3f%16.3e\n",
          tout[it],z[iz],u2[iz,it]));
    }
    cat(sprintf("\n       t        z           u3(r=0,z,t)\n"));
    for(iz in ivz){
      cat(sprintf("%7.1f%9.3f%16.3e\n",
          tout[it],z[iz],u3[iz,it]));
    }
    cat(sprintf("\n       t        z           u4(r=0,z,t)\n"));
    for(iz in ivz){
      cat(sprintf("%7.1f%9.3f%16.3e\n",
          tout[it],z[iz],u4[iz,it]));
    }
    cat(sprintf("\n       t        z           u5(r=0,z,t)\n"));
    for(iz in ivz){
      cat(sprintf("%7.1f%9.3f%16.3e\n",
          tout[it],z[iz],u5[iz,it]));
    }
    cat(sprintf("\n       t        z           u6(r=0,z,t)\n"));
    for(iz in ivz){
      cat(sprintf("%7.1f%9.3f%16.3e\n",
          tout[it],z[iz],u6[iz,it]));
    }
  }
#
# Calls to ODE routine
  cat(sprintf("\n\n ncall = %5d\n\n",ncall));
#
# Plot PDE solutions
#
# u1m(z,t)
# 2D
  par(mfrow=c(1,1));
```

```
  matplot(x=z[2:nz],y=u1m[2:nz,],type="l",xlab="z",
    ylab="u1m(r=0,z,t)",lty=1,main="",lwd=2,col="black");
#
# 3D
  persp(z,tout,u1m,theta=45,phi=30,xlim=c(zl,zu),
    ylim=c(t0,tf),xlab="z",ylab="t",zlab="u1m(r=0,z,t)");
#
# u2m(z,t)
# 2D
  par(mfrow=c(1,1));
  matplot(x=z[2:nz],y=u2m[2:nz,],type="l",xlab="z",
    ylab="u2m(r=0,z,t)",lty=1,main="",lwd=2,col="black");
#
# 3D
  persp(z,tout,u2m,theta=45,phi=30,xlim=c(zl,zu),
    ylim=c(t0,tf),xlab="z",ylab="t",zlab="u2m(r=0,z,t)");
#
# u1(z,t)
# 2D
  par(mfrow=c(1,1));
  matplot(x=z[2:nz],y=u1[2:nz,],type="l",xlab="z",
    ylab="u1(r=0,z,t)",lty=1,main="",lwd=2,col="black");
#
# 3D
  persp(z,tout,u1,theta=45,phi=30,xlim=c(zl,zu),
    ylim=c(t0,tf),xlab="z",ylab="t",zlab="u1(r=0,z,t)");
#
# u2(z,t)
# 2D
  par(mfrow=c(1,1));
  matplot(x=z[2:nz],y=u2[2:nz,],type="l",xlab="z",
    ylab="u2(r=0,z,t)",lty=1,main="",lwd=2,col="black");
#
# 3D
  persp(z,tout,u2,theta=45,phi=30,xlim=c(zl,zu),
    ylim=c(t0,tf),xlab="z",ylab="t",zlab="u2(r=0,z,t)");
#
# u3(z,t)
# 2D
  par(mfrow=c(1,1));
  matplot(x=z[2:nz],y=u3[2:nz,],type="l",xlab="z",
    ylab="u3(r=0,z,t)",lty=1,main="",lwd=2,col="black");
```

```
#
# 3D
  persp(z,tout,u3,theta=45,phi=30,xlim=c(zl,zu),
    ylim=c(t0,tf),xlab="z",ylab="t",zlab="u3(r=0,z,t)");
#
# u4(z,t)
# 2D
  par(mfrow=c(1,1));
  matplot(x=z[2:nz],y=u4[2:nz,],type="l",xlab="z",
    ylab="u4(r=0,z,t)",lty=1,main="",lwd=2,col="black");
#
# 3D
  persp(z,tout,u4,theta=45,phi=30,xlim=c(zl,zu),
    ylim=c(t0,tf),xlab="z",ylab="t",zlab="u4(r=0,z,t)");
#
# u5(z,t)
# 2D
  par(mfrow=c(1,1));
  matplot(x=z[2:nz],y=u5[2:nz,],type="l",xlab="z",
    ylab="u5(r=0,z,t)",lty=1,main="",lwd=2,col="black");
#
# 3D
  persp(z,tout,u5,theta=45,phi=30,xlim=c(zl,zu),
    ylim=c(t0,tf),xlab="z",ylab="t",zlab="u5(r=0,z,t)");
#
# u6(z,t)
# 2D
  par(mfrow=c(1,1));
  matplot(x=z[2:nz],y=u6[2:nz,],type="l",xlab="z",
    ylab="u6(r=0,z,t)",lty=1,main="",lwd=2,col="black");
#
# 3D
  persp(z,tout,u6,theta=45,phi=30,xlim=c(zl,zu),
    ylim=c(t0,tf),xlab="z",ylab="t",zlab="u6(r=0,z,t)");
```

Listing 5.1: Main program for eqs. (5.1), (5.4).

We can note the following details of Listing 5.1 (with some repetition of the discussion of Listing 4.1 so that this explanation is self contained).

- Previous workspaces are deleted.

```
#
# 2D TC model
```

```
#
# Delete previous workspaces
  rm(list=ls(all=TRUE))
```

- The R ODE integrator library deSolve is accessed [2].

```
#
# Access ODE integrator
  library("deSolve");
#
# Access functions for numerical solution
  setwd("f:/tissue engineering/chap5");
  source("pde2b.R");
```

Then the directory with the files for the solution of eqs. (5.1) to (5.10) is designated. Note that setwd (set working directory) uses / rather than the usual \. The ordinary differential equation/method of lines (ODE/MOL)[1] routine, pde2b.R, is accessed through setwd.

- The model parameters are specified numerically.

```
#
# Parameters
#
# u1m
  D1mr=0.5;
  D1mz=0.5;
  p1=1;
  Qm1=1;
  cm1=1;
  pQm1=p1*Qm1;
  u1mb=1;
  u1m0=0;
#
# u2m
  D2mr=0.5;
  D2mz=0.5;
  p2=1;
  Qm2=1;
  cm2=1;
  pQm2=p2*Qm2;
```

[1] The method of lines is a general numerical algorithm for PDEs in which the boundary value (spatial) derivatives are replaced with algebraic approximations, in this case, finite differences (FDs). The resulting system of initial value ODEs is then integrated (solved) with a library ODE integrator.

```
    u2mb=1;
    u2m0=0;
#
# u1
    D1r=0.25;
    D1z=0.25;
    a11=1;
    u1b=1;
    u10=0;
#
# u2
    D2r=0.25;
    D2z=0.25;
    a21=1;a22=1;
    u2b=0;
    u20=0;
#
# u3
    D3r=0.25;
    D3z=0.25;
    a31=1;a32=1;
    u3b=0;
    u30=0;
#
# u4
    D4r=0.25;
    D4z=0.25;
    a41=1;a42=1;
    a43=1;a44=1;
    a45=1;
    u4b=0;
    u40=0;
#
# u5
    D5r=0.25;
    D5z=0.25;
    a51=1;a52=1;
    a53=1;a54=1;
    a55=1;
    u5b=0;
    u50=0;
#
```

```
# u6
  D6r=0.25;
  D6z=0.25;
  a61=1;a62=1;
  a63=1;
  u6b=0;
  u60=0;
#
# Parameters probability functions
  p0m=1;p1m=0.1;
   g4=1; g5=1;
    m=2;   n=2;
```

For example,

```
#
# Parameters
#
# u1m
  D1mr=0.5;
  D1mz=0.5;
  p1=1;
  Qm1=1;
  cm1=1;
  pQm1=p1*Qm1;
  u1mb=1;
  u1m0=0;
```

are the parameters for eqs. (5.1-1), (5.1-3), (5.2), where
- D1mr, D1mz: Effective O_2 diffusivities in eq. (5.1-1).
- p1,Qm1,cm1: Parameters for q_1 of eq. (5.1-3).
- u1mb: Boundary condition (BC) at $z = z_l$, $g_{1m}(r, t)$ eq. (5.2-4).
- u1m0: Initial condition (IC), $f_{1m}(r, z)$, eq. (5.2-1).
- A spatial grid in r for eqs. (5.1) to (5.10) is defined with 6 points so that r = 0,1/5=0.2, ..., 1. The stem cell radius is a normalized value, $r = r_u = 1$.

```
#
# Spatial grid (in r)
  nr=6;rl=0;ru=1;dr=(ru-rl)/(nr-1);dr2=dr^2;
  r=seq(from=rl,to=ru,by=dr);
```

The low value nr=6 was selected to limit the number of spatial grid points to a manageable level while achieving acceptable accuracy over the half-width in r.

- A spatial grid in z for eqs. (5.1) to (5.10) is defined with 21 points so that z = 0,1/20= 0.05,...,1. The stem cell length is a normalized value, $z = z_u = 1$.

```
#
# Spatial grid (in z)
  nz=21;zl=0;zu=1;dz=(zu-zl)/(nz-1);dz2=dz^2;
  z=seq(from=zl,to=zu,by=dz);
```

The value nz=21 was selected large enough to avoid gridding (nonsmooth) effects in z (as reflected in the graphical output that follows).
- An interval in t is defined for 11 output points, so that tout=0,1/10=0.1,...,1. The time scale is normalized with $t_f = 1$ specified as the final time that is considered appropriate, e.g., minutes, days.

```
#
# Independent variable for ODE integration
  t0=0;tf=1;nout=11;
  tout=seq(from=t0,to=tf,by=(tf-t0)/(nout-1));
```

- ICs (5.2-1), (5.3-1), (5.5-1), (5.6-1), ..., (5.10-1) are implemented.

```
#
# Initial conditions (t=0)
  u0=rep(0,8*nr*nz);
  for(iz in 1:nz){
  for(ir in 1:nr){
    u0[(iz-1)*nr+ir]          =u1m0;
    u0[(iz-1)*nr+ir+nr*nz]    =u2m0;
    u0[(iz-1)*nr+ir+2*nr*nz]= u10;
    u0[(iz-1)*nr+ir+3*nr*nz]= u20;
    u0[(iz-1)*nr+ir+4*nr*nz]= u30;
    u0[(iz-1)*nr+ir+5*nr*nz]= u40;
    u0[(iz-1)*nr+ir+6*nr*nz]= u50;
    u0[(iz-1)*nr+ir+7*nr*nz]= u60;
  }
  }
  ncall=0;
```

Also, the counter for the calls to pde2b is initialized.
- The system of nz=8*(6*21)=1008 ODEs is integrated by the library integrator lsodes (available in deSolve, [2]). As expected, the inputs to lsodes are the ODE function, pde2b, the IC vector u0, and the vector of output values of t, tout. The length of u0 (1008) informs lsodes how many ODEs are to be integrated. func,y,times are reserved names.

```
#
# ODE integration
  out=lsodes(y=u0,times=tout,func=pde2b,
      sparsetype ="sparseint",rtol=1e-6,
      atol=1e-6,maxord=5);
  nrow(out)
  ncol(out)
```

`nrow,ncol` confirm the dimensions of `out`.

- Arrays are defined for the eight PDE dependent variables returned in `out` from `lsodes`.

```
#
# Arrays for displaying, plotting numerical
# solutions
  u1m=matrix(0,nrow=nz,ncol=nout);
  u2m=matrix(0,nrow=nz,ncol=nout);
   u1=matrix(0,nrow=nz,ncol=nout);
   u2=matrix(0,nrow=nz,ncol=nout);
   u3=matrix(0,nrow=nz,ncol=nout);
   u4=matrix(0,nrow=nz,ncol=nout);
   u5=matrix(0,nrow=nz,ncol=nout);
   u6=matrix(0,nrow=nz,ncol=nout);
```

- The eight solutions $u_{1m}(r, z, t)$ to $u_6(r, z, t)$ are placed in the previously defined arrays.

```
#
# Save solutions from lsodes
  for(it in 1:nout){
  for(iz in 1:nz){
      u1m[iz,it]=out[it,(iz-1)*nr+ir+1];
      u1m[1,it]=u1mb;
      u2m[iz,it]=out[it,(iz-1)*nr+ir+1+ nr*nz];
      u2m[1,it]=u2mb;
      u1[iz,it]=out[it,(iz-1)*nr+ir+1+2*nr*nz];
      u1[1,it]=u1b;
      u2[iz,it]=out[it,(iz-1)*nr+ir+1+3*nr*nz];
      u2[1,it]=u2b;
      u3[iz,it]=out[it,(iz-1)*nr+ir+1+4*nr*nz];
      u3[1,it]=u3b;
      u4[iz,it]=out[it,(iz-1)*nr+ir+1+5*nr*nz];
      u4[1,it]=u4b;
      u5[iz,it]=out[it,(iz-1)*nr+ir+1+6*nr*nz];
      u5[1,it]=u5b;
```

```
      u6[iz,it]=out[it,(iz-1)*nr+ir+1+7*nr*nz];
      u6[1,it]=u6b;
  }
  }
```

These solutions include the BCs at $z = z_l$, for example, BC (5.2-4) is programmed as
u1m[1,it]=u1mb;.

The offset +1 is required because the first element of the solution vectors in out is the
value of t and the 2 to 1009 elements are the 1008 values of u_{1m} to u_6. These dimensions
from the preceding calls to nrow,ncol are confirmed in the subsequent output.

• The eight solutions for $r = 0$ (ir=1) are displayed. Every fifth value in t and z is selected
with the two by=5.

```
#
# Display solutions
  iv=seq(from=1,to=nout,by=5);
  for(it in iv){
  ivz=seq(from=1,to=nz,by=5);
    ir=1;
    cat(sprintf("\n      t        z          u1m(r=0,z,t)\n"));
    for(iz in ivz){
      cat(sprintf("%7.1f%9.3f%16.3e\n",
          tout[it],z[iz],u1m[iz,it]));
    }
    cat(sprintf("\n      t        z          u2m(r=0,z,t)\n"));
    for(iz in ivz){
      cat(sprintf("%7.1f%9.3f%16.3e\n",
          tout[it],z[iz],u2m[iz,it]));
    }
    cat(sprintf("\n      t        z          u1(r=0,z,t)\n"));
    for(iz in ivz){
      cat(sprintf("%7.1f%9.3f%16.3e\n",
          tout[it],z[iz],u1[iz,it]));
    }
    cat(sprintf("\n      t        z          u2(r=0,z,t)\n"));
    for(iz in ivz){
      cat(sprintf("%7.1f%9.3f%16.3e\n",
          tout[it],z[iz],u2[iz,it]));
    }
    cat(sprintf("\n      t        z          u3(r=0,z,t)\n"));
    for(iz in ivz){
      cat(sprintf("%7.1f%9.3f%16.3e\n",
          tout[it],z[iz],u3[iz,it]));
```

```
    }
    cat(sprintf("\n      t       z          u4(r=0,z,t)\n"));
    for(iz in ivz){
      cat(sprintf("%7.1f%9.3f%16.3e\n",
        tout[it],z[iz],u4[iz,it]));
    }
    cat(sprintf("\n      t       z          u5(r=0,z,t)\n"));
    for(iz in ivz){
      cat(sprintf("%7.1f%9.3f%16.3e\n",
        tout[it],z[iz],u5[iz,it]));
    }
    cat(sprintf("\n      t       z          u6(r=0,z,t)\n"));
    for(iz in ivz){
      cat(sprintf("%7.1f%9.3f%16.3e\n",
        tout[it],z[iz],u6[iz,it]));
    }
  }
```

• The number of calls to pde2b is displayed at the end of the solution.

```
#
# Calls to ODE routine
  cat(sprintf("\n\n ncall = %5d\n\n",ncall));
```

• The eight PDE solutions are plotted in 2D with the matplot utility, and in 3D with the persp utility. For the 2D plots, the solutions at $z = 0$ are not included to avoid the discontinuities between the ICs and BCs. For example, for u_{1m}, x=z[2:nz],y=u1m[2:nz,] does not include $z = z_l = 0$. Including the solutions at $z = z_l = 0$ is left as an exercise. par(mfrow=c(1,1)); specifies a 1×1 matrix of plots, that is, one plot per page.

```
#
# Plot PDE solutions
#
# u1m(z,t)
# 2D
  par(mfrow=c(1,1));
  matplot(x=z[2:nz],y=u1m[2:nz,],type="l",xlab="z",
    ylab="u1m(r=0,z,t)",lty=1,main="",lwd=2,col="black");
#
# 3D
  persp(z,tout,u1m,theta=45,phi=30,xlim=c(zl,zu),
    ylim=c(t0,tf),xlab="z",ylab="t",zlab="u1m(r=0,z,t)");
#
```

```
# u2m(z,t)
# 2D
  par(mfrow=c(1,1));
  matplot(x=z[2:nz],y=u2m[2:nz,],type="l",xlab="z",
    ylab="u2m(r=0,z,t)",lty=1,main="",lwd=2,col="black");
#
# 3D
  persp(z,tout,u2m,theta=45,phi=30,xlim=c(zl,zu),
    ylim=c(t0,tf),xlab="z",ylab="t",zlab="u2m(r=0,z,t)");
#
# u1(z,t)
# 2D
  par(mfrow=c(1,1));
  matplot(x=z[2:nz],y=u1[2:nz,],type="l",xlab="z",
    ylab="u1(r=0,z,t)",lty=1,main="",lwd=2,col="black");
#
# 3D
  persp(z,tout,u1,theta=45,phi=30,xlim=c(zl,zu),
    ylim=c(t0,tf),xlab="z",ylab="t",zlab="u1(r=0,z,t)");
#
# u2(z,t)
# 2D
  par(mfrow=c(1,1));
  matplot(x=z[2:nz],y=u2[2:nz,],type="l",xlab="z",
    ylab="u2(r=0,z,t)",lty=1,main="",lwd=2,col="black");
#
# 3D
  persp(z,tout,u2,theta=45,phi=30,xlim=c(zl,zu),
    ylim=c(t0,tf),xlab="z",ylab="t",zlab="u2(r=0,z,t)");
#
# u3(z,t)
# 2D
  par(mfrow=c(1,1));
  matplot(x=z[2:nz],y=u3[2:nz,],type="l",xlab="z",
    ylab="u3(r=0,z,t)",lty=1,main="",lwd=2,col="black");
#
# 3D
  persp(z,tout,u3,theta=45,phi=30,xlim=c(zl,zu),
    ylim=c(t0,tf),xlab="z",ylab="t",zlab="u3(r=0,z,t)");
#
# u4(z,t)
# 2D
```

```
   par(mfrow=c(1,1));
   matplot(x=z[2:nz],y=u4[2:nz,],type="l",xlab="z",
     ylab="u4(r=0,z,t)",lty=1,main="",lwd=2,col="black");
#
# 3D
   persp(z,tout,u4,theta=45,phi=30,xlim=c(zl,zu),
     ylim=c(t0,tf),xlab="z",ylab="t",zlab="u4(r=0,z,t)");
#
# u5(z,t)
# 2D
   par(mfrow=c(1,1));
   matplot(x=z[2:nz],y=u5[2:nz,],type="l",xlab="z",
     ylab="u5(r=0,z,t)",lty=1,main="",lwd=2,col="black");
#
# 3D
   persp(z,tout,u5,theta=45,phi=30,xlim=c(zl,zu),
     ylim=c(t0,tf),xlab="z",ylab="t",zlab="u5(r=0,z,t)");
#
# u6(z,t)
# 2D
   par(mfrow=c(1,1));
   matplot(x=z[2:nz],y=u6[2:nz,],type="l",xlab="z",
     ylab="u6(r=0,z,t)",lty=1,main="",lwd=2,col="black");
#
# 3D
   persp(z,tout,u6,theta=45,phi=30,xlim=c(zl,zu),
     ylim=c(t0,tf),xlab="z",ylab="t",zlab="u6(r=0,z,t)");
```

This completes the discussion of the main program of Listing 5.1. The ODE/PDE routine pde2b called by lsodes is considered next.

5.3.2 ODE/MOL routine

pde2b called by lsodes in Listing 5.1 follows.

```
   pde2b=function(t,u,parm){
#
# Function pde2b computes the t derivative
# of u1m(r,z,t), u2m(r,z,t), u1(r,z,t),
#     u2(r,z,t),  u3(r,z,t), u4(r,z,t),
#     u5(r,z,t),  u6(r,z,t)
#
# 1D vector to 2D matrices
```

```
  u1m=matrix(0,nrow=nr,ncol=nz);
  u2m=matrix(0,nrow=nr,ncol=nz);
   u1=matrix(0,nrow=nr,ncol=nz);
   u2=matrix(0,nrow=nr,ncol=nz);
   u3=matrix(0,nrow=nr,ncol=nz);
   u4=matrix(0,nrow=nr,ncol=nz);
   u5=matrix(0,nrow=nr,ncol=nz);
   u6=matrix(0,nrow=nr,ncol=nz);
  for(iz in 1:nz){
  for(ir in 1:nr){
    u1m[ir,iz]=u[(iz-1)*nr+ir];
    u2m[ir,iz]=u[(iz-1)*nr+ir+  nr*nz];
     u1[ir,iz]=u[(iz-1)*nr+ir+2*nr*nz];
     u2[ir,iz]=u[(iz-1)*nr+ir+3*nr*nz];
     u3[ir,iz]=u[(iz-1)*nr+ir+4*nr*nz];
     u4[ir,iz]=u[(iz-1)*nr+ir+5*nr*nz];
     u5[ir,iz]=u[(iz-1)*nr+ir+6*nr*nz];
     u6[ir,iz]=u[(iz-1)*nr+ir+7*nr*nz];
  }
  }
#
# u1mrr, Neumann BC, r=0,ru
  u1mrr=matrix(0,nrow=nr,ncol=nz);
  for(iz in 1:nz){
  for(ir in 1:nr){
    if(ir==1){
      u1mrr[1,iz]=4*(u1m[2,iz]-u1m[1,iz])/dr2;}
    if(ir==nr){
      u1mrr[nr,iz]=2*(u1m[nr-1,iz]-u1m[nr,iz])/dr2;}
    if((ir>1)&(ir<nr)){
      u1mrr[ir,iz]=(u1m[ir+1,iz]-2*u1m[ir,iz]+u1m[ir-1,iz])/dr2+
                (1/r[ir])*(u1m[ir+1,iz]-u1m[ir-1,iz])/(2*dr);}
  }
  }
#
# u1mzz, Dirichlet, z=zl, Neumann BC, z=zu
  u1mzz=matrix(0,nrow=nr,ncol=nz);
  for(iz in 1:nz){
  for(ir in 1:nr){
    if(iz==1){
      u1m[ir,1]=u1mb};
    if(iz==nz){
```

```
              u1mzz[ir,nz]=2*(u1m[ir,nz-1]-u1m[ir,nz])/dz2;}
          if((iz>1)&(iz<nz)){
              u1mzz[ir,iz]=(u1m[ir,iz+1]-2*u1m[ir,iz]+u1m[ir,iz-1])/dz2;}
        }
        }
  #
  # u2mrr, Neumann BC, r=0,ru
      u2mrr=matrix(0,nrow=nr,ncol=nz);
      for(iz in 1:nz){
      for(ir in 1:nr){
        if(ir==1){
          u2mrr[1,iz]=4*(u2m[2,iz]-u2m[1,iz])/dr2;}
        if(ir==nr){
          u2mrr[nr,iz]=2*(u2m[nr-1,iz]-u2m[nr,iz])/dr2;}
        if((ir>1)&(ir<nr)){
          u2mrr[ir,iz]=(u2m[ir+1,iz]-2*u2m[ir,iz]+u2m[ir-1,iz])/dr2+
                      (1/r[ir])*(u2m[ir+1,iz]-u2m[ir-1,iz])/(2*dr);}
        }
        }
  #
  # u2mzz, Dirichlet, z=zl, Neumann BC, z=zu
      u2mzz=matrix(0,nrow=nr,ncol=nz);
      for(iz in 1:nz){
      for(ir in 1:nr){
        if(iz==1){
          u2m[ir,1]=u2mb};
        if(iz==nz){
          u2mzz[ir,nz]=2*(u2m[ir,nz-1]-u2m[ir,nz])/dz2;}
        if((iz>1)&(iz<nz)){
          u2mzz[ir,iz]=(u2m[ir,iz+1]-2*u2m[ir,iz]+u2m[ir,iz-1])/dz2;}
        }
        }
  #
  # u1rr, Neumann BC, r=0,ru
      u1rr=matrix(0,nrow=nr,ncol=nz);
      for(iz in 1:nz){
      for(ir in 1:nr){
        if(ir==1){
          u1rr[1,iz]=4*(u1[2,iz]-u1[1,iz])/dr2;}
        if(ir==nr){
          u1rr[nr,iz]=2*(u1[nr-1,iz]-u1[nr,iz])/dr2;}
        if((ir>1)&(ir<nr)){
```

```
        u1rr[ir,iz]=(u1[ir+1,iz]-2*u1[ir,iz]+u1[ir-1,iz])/dr2+
                 (1/r[ir])*(u1[ir+1,iz]-u1[ir-1,iz])/(2*dr);}
  }
  }
#
# u1zz, Dirichlet, z=zl, Neumann BC, z=zu
  u1zz=matrix(0,nrow=nr,ncol=nz);
  for(iz in 1:nz){
  for(ir in 1:nr){
    if(iz==1){
      u1[ir,1]=u1b};
    if(iz==nz){
      u1zz[ir,nz]=2*(u1[ir,nz-1]-u1[ir,nz])/dz2;}
    if((iz>1)&(iz<nz)){
      u1zz[ir,iz]=(u1[ir,iz+1]-2*u1[ir,iz]+u1[ir,iz-1])/dz2;}
  }
  }
#
# u2rr, Neumann BC, r=0,ru
  u2rr=matrix(0,nrow=nr,ncol=nz);
  for(iz in 1:nz){
  for(ir in 1:nr){
    if(ir==1){
      u2rr[1,iz]=4*(u2[2,iz]-u2[1,iz])/dr2;}
    if(ir==nr){
      u2rr[nr,iz]=2*(u2[nr-1,iz]-u2[nr,iz])/dr2;}
    if((ir>1)&(ir<nr)){
      u2rr[ir,iz]=(u2[ir+1,iz]-2*u2[ir,iz]+u2[ir-1,iz])/dr2+
                 (1/r[ir])*(u2[ir+1,iz]-u2[ir-1,iz])/(2*dr);}
  }
  }
#
# u2zz, Dirichlet, z=zl, Neumann BC, z=zu
  u2zz=matrix(0,nrow=nr,ncol=nz);
  for(iz in 1:nz){
  for(ir in 1:nr){
    if(iz==1){
      u2[ir,1]=u2b};
    if(iz==nz){
      u2zz[ir,nz]=2*(u2[ir,nz-1]-u2[ir,nz])/dz2;}
    if((iz>1)&(iz<nz)){
      u2zz[ir,iz]=(u2[ir,iz+1]-2*u2[ir,iz]+u2[ir,iz-1])/dz2;}
```

```
    }
    }
#
# u3rr, Neumann BC, r=0,ru
  u3rr=matrix(0,nrow=nr,ncol=nz);
  for(iz in 1:nz){
  for(ir in 1:nr){
    if(ir==1){
      u3rr[1,iz]=4*(u3[2,iz]-u3[1,iz])/dr2;}
    if(ir==nr){
      u3rr[nr,iz]=2*(u3[nr-1,iz]-u3[nr,iz])/dr2;}
    if((ir>1)&(ir<nr)){
      u3rr[ir,iz]=(u3[ir+1,iz]-2*u3[ir,iz]+u3[ir-1,iz])/dr2+
                 (1/r[ir])*(u3[ir+1,iz]-u3[ir-1,iz])/(2*dr);}
  }
  }
#
# u3zz, Dirichlet, z=zl, Neumann BC, z=zu
  u3zz=matrix(0,nrow=nr,ncol=nz);
  for(iz in 1:nz){
  for(ir in 1:nr){
    if(iz==1){
      u3[ir,1]=u3b};
    if(iz==nz){
      u3zz[ir,nz]=2*(u3[ir,nz-1]-u3[ir,nz])/dz2;}
    if((iz>1)&(iz<nz)){
      u3zz[ir,iz]=(u3[ir,iz+1]-2*u3[ir,iz]+u3[ir,iz-1])/dz2;}
  }
  }
#
# u4rr, Neumann BC, r=0,ru
  u4rr=matrix(0,nrow=nr,ncol=nz);
  for(iz in 1:nz){
  for(ir in 1:nr){
    if(ir==1){
      u4rr[1,iz]=4*(u4[2,iz]-u4[1,iz])/dr2;}
    if(ir==nr){
      u4rr[nr,iz]=2*(u4[nr-1,iz]-u4[nr,iz])/dr2;}
    if((ir>1)&(ir<nr)){
      u4rr[ir,iz]=(u4[ir+1,iz]-2*u4[ir,iz]+u4[ir-1,iz])/dr2+
                 (1/r[ir])*(u4[ir+1,iz]-u4[ir-1,iz])/(2*dr);}
  }
```

```
    }
#
# u4zz, Dirichlet, z=zl, Neumann BC, z=zu
  u4zz=matrix(0,nrow=nr,ncol=nz);
  for(iz in 1:nz){
  for(ir in 1:nr){
    if(iz==1){
      u4[ir,1]=u4b};
    if(iz==nz){
      u4zz[ir,nz]=2*(u4[ir,nz-1]-u4[ir,nz])/dz2;}
    if((iz>1)&(iz<nz)){
      u4zz[ir,iz]=(u4[ir,iz+1]-2*u4[ir,iz]+u4[ir,iz-1])/dz2;}
  }
  }
#
# u5rr, Neumann BC, r=0,ru
  u5rr=matrix(0,nrow=nr,ncol=nz);
  for(iz in 1:nz){
  for(ir in 1:nr){
    if(ir==1){
      u5rr[1,iz]=4*(u5[2,iz]-u5[1,iz])/dr2;}
    if(ir==nr){
      u5rr[nr,iz]=2*(u5[nr-1,iz]-u5[nr,iz])/dr2;}
    if((ir>1)&(ir<nr)){
      u5rr[ir,iz]=(u5[ir+1,iz]-2*u5[ir,iz]+u5[ir-1,iz])/dr2+
                  (1/r[ir])*(u5[ir+1,iz]-u5[ir-1,iz])/(2*dr);}
  }
  }
#
# u5zz, Dirichlet, z=zl, Neumann BC, z=zu
  u5zz=matrix(0,nrow=nr,ncol=nz);
  for(iz in 1:nz){
  for(ir in 1:nr){
    if(iz==1){
      u5[ir,1]=u5b};
    if(iz==nz){
      u5zz[ir,nz]=2*(u5[ir,nz-1]-u5[ir,nz])/dz2;}
    if((iz>1)&(iz<nz)){
      u5zz[ir,iz]=(u5[ir,iz+1]-2*u5[ir,iz]+u5[ir,iz-1])/dz2;}
  }
  }
#
```

```
# u6rr, Neumann BC, r=0,ru
  u6rr=matrix(0,nrow=nr,ncol=nz);
  for(iz in 1:nz){
  for(ir in 1:nr){
    if(ir==1){
      u6rr[1,iz]=4*(u6[2,iz]-u6[1,iz])/dr2;}
    if(ir==nr){
      u6rr[nr,iz]=2*(u6[nr-1,iz]-u6[nr,iz])/dr2;}
    if((ir>1)&(ir<nr)){
      u6rr[ir,iz]=(u6[ir+1,iz]-2*u6[ir,iz]+u6[ir-1,iz])/dr2+
                  (1/r[ir])*(u6[ir+1,iz]-u6[ir-1,iz])/(2*dr);}
  }
  }
#
# u6zz, Dirichlet, z=zl, Neumann BC, z=zu
  u6zz=matrix(0,nrow=nr,ncol=nz);
  for(iz in 1:nz){
  for(ir in 1:nr){
    if(iz==1){
      u6[ir,1]=u6b};
    if(iz==nz){
      u6zz[ir,nz]=2*(u6[ir,nz-1]-u6[ir,nz])/dz2;}
    if((iz>1)&(iz<nz)){
      u6zz[ir,iz]=(u6[ir,iz+1]-2*u6[ir,iz]+u6[ir,iz-1])/dz2;}
  }
  }
#
# u1mt, u2mt, u1t, u2t, u3t, u4t, u5t, u6t
  u1mt=matrix(0,nrow=nr,ncol=nz);
  u2mt=matrix(0,nrow=nr,ncol=nz);
   u1t=matrix(0,nrow=nr,ncol=nz);
   u2t=matrix(0,nrow=nr,ncol=nz);
   u3t=matrix(0,nrow=nr,ncol=nz);
   u4t=matrix(0,nrow=nr,ncol=nz);
   u5t=matrix(0,nrow=nr,ncol=nz);
   u6t=matrix(0,nrow=nr,ncol=nz);
  for(iz in 1:nz){
  for(ir in 1:nr){
#
#   Michaelis-Menten, probability functions
    q1=pQm1*u1m[ir,iz]/(cm1+u1m[ir,iz])*u2m[ir,iz];
    q2=pQm2*u2m[ir,iz]/(cm2+u2m[ir,iz])*u1m[ir,iz];
```

```
      p0=p0m/(1+(g4*u4[ir,iz])^m);
      p1=p1m/(1+(g5*u5[ir,iz])^n);
#
#     PDEs
      u1mt[ir,iz]=D1mr*u1mrr[ir,iz]+D1mz*u1mzz[ir,iz]-q1;
      u2mt[ir,iz]=D2mr*u2mrr[ir,iz]+D2mz*u2mzz[ir,iz]-q2;
       u1t[ir,iz]=D1r*u1rr[ir,iz]+D1z*u1zz[ir,iz]+
                    a11*(2*p0-1)*u1[ir,iz]+q1;
       u2t[ir,iz]=D2r*u2rr[ir,iz]+D2z*u2zz[ir,iz]+
                    a21*(2*(1-p0)*u1[ir,iz])+
                    a22*((2*p1-1)*u2[ir,iz]);
       u3t[ir,iz]=D3r*u3rr[ir,iz]+D3z*u3zz[ir,iz]+
                    a31*(2*(1-p1)*u2[ir,iz])-
                    a32*u3[ir,iz];
       u4t[ir,iz]=D4r*u4rr[ir,iz]+D4z*u4zz[ir,iz]+
                    a41*u1[ir,iz]+a42*u2[ir,iz]+
                    a43*u3[ir,iz]-a44*u4[ir,iz]-
                    a45*u4[ir,iz]*u6[ir,iz];
       u5t[ir,iz]=D5r*u5rr[ir,iz]+D5z*u5zz[ir,iz]+
                    a51*u1[ir,iz]+a52*u2[ir,iz]+
                    a53*u3[ir,iz]-a54*u5[ir,iz]-
                    a55*u5[ir,iz]*u6[ir,iz];
       u6t[ir,iz]=D6r*u6rr[ir,iz]+D6z*u6zz[ir,iz]-
                    a61*u6[ir,iz]-a62*u4[ir,iz]*u6[ir,iz]-
                    a63*u5[ir,iz]*u6[ir,iz];
     u1mt[ir,1]=0;
     u2mt[ir,1]=0;
      u1t[ir,1]=0;
      u2t[ir,1]=0;
      u3t[ir,1]=0;
      u4t[ir,1]=0;
      u5t[ir,1]=0;
      u6t[ir,1]=0;
   }
   }
#
# 2D matrices to 1D vector
  ut=rep(0,8*nr*nz)
  for(iz in 1:nz){
  for(ir in 1:nr){
    ut[(iz-1)*nr+ir]          =u1mt[ir,iz];
    ut[(iz-1)*nr+ir+nr*nz] =u2mt[ir,iz];
```

```
    ut[(iz-1)*nr+ir+2*nr*nz]=u1t[ir,iz];
    ut[(iz-1)*nr+ir+3*nr*nz]=u2t[ir,iz];
    ut[(iz-1)*nr+ir+4*nr*nz]=u3t[ir,iz];
    ut[(iz-1)*nr+ir+5*nr*nz]=u4t[ir,iz];
    ut[(iz-1)*nr+ir+6*nr*nz]=u5t[ir,iz];
    ut[(iz-1)*nr+ir+7*nr*nz]=u6t[ir,iz];
  }
  }
#
# Increment calls to pde2b
  ncall <<- ncall+1;
#
# Return derivative vector
  return(list(c(ut)));
  }
```

Listing 5.2: ODE/MOL routine pde2b for eqs. (5.1), (5.4).

We can note the following details of Listing 5.2 (with some repetition of the discussion of Listing 4.2 so that this explanation is self contained).

- The function is defined.

```
  pde2b=function(t,u,parm){
#
# Function pde2b computes the t derivative
# of u1m(r,z,t), u2m(r,z,t), u1(r,z,t),
#      u2(r,z,t),  u3(r,z,t), u4(r,z,t),
#      u5(r,z,t),  u6(r,z,t)
```

  t is the current value of t in eqs. (5.1), (5.4). u is the 1008-vector of ODE/PDE dependent variables. parm is an argument to pass parameters to pde2b (unused, but required in the argument list). The arguments must be listed in the order stated to properly interface with lsodes called in the main program of Listing 5.1. The derivative vector of the LHS of eqs. (5.1), (5.4) is calculated and returned to lsodes as explained subsequently.

- The input vector u is placed in matrices u1m, u2m, ..., u6 to facilitate the programming of eqs. (5.1), (5.4).

- #
```
# 1D vector to 2D matrices
  u1m=matrix(0,nrow=nr,ncol=nz);
  u2m=matrix(0,nrow=nr,ncol=nz);
   u1=matrix(0,nrow=nr,ncol=nz);
   u2=matrix(0,nrow=nr,ncol=nz);
   u3=matrix(0,nrow=nr,ncol=nz);
```

```
   u4=matrix(0,nrow=nr,ncol=nz);
   u5=matrix(0,nrow=nr,ncol=nz);
   u6=matrix(0,nrow=nr,ncol=nz);
 for(iz in 1:nz){
 for(ir in 1:nr){
   u1m[ir,iz]=u[(iz-1)*nr+ir];
   u2m[ir,iz]=u[(iz-1)*nr+ir+  nr*nz];
   u1[ir,iz]=u[(iz-1)*nr+ir+2*nr*nz];
   u2[ir,iz]=u[(iz-1)*nr+ir+3*nr*nz];
   u3[ir,iz]=u[(iz-1)*nr+ir+4*nr*nz];
   u4[ir,iz]=u[(iz-1)*nr+ir+5*nr*nz];
   u5[ir,iz]=u[(iz-1)*nr+ir+6*nr*nz];
   u6[ir,iz]=u[(iz-1)*nr+ir+7*nr*nz];
 }
 }
```

- The radial group

$$\frac{\partial^2 u_{1m}}{\partial r^2} + \frac{1}{r}\frac{\partial u_{1m}}{\partial r}$$

in eq. (5.1-1) is computed as u1mrr with finite differences as explained for eq. (4.1-1).

```
#
# u1mrr, Neumann BC, r=0,ru
  u1mrr=matrix(0,nrow=nr,ncol=nz);
  for(iz in 1:nz){
  for(ir in 1:nr){
    if(ir==1){
      u1mrr[1,iz]=4*(u1m[2,iz]-u1m[1,iz])/dr2;}
    if(ir==nr){
      u1mrr[nr,iz]=2*(u1m[nr-1,iz]-u1m[nr,iz])/dr2;}
    if((ir>1)&(ir<nr)){
      u1mrr[ir,iz]=(u1m[ir+1,iz]-2*u1m[ir,iz]+u1m[ir-1,iz])/dr2+
                   (1/r[ir])*(u1m[ir+1,iz]-u1m[ir-1,iz])/(2*dr);}
  }
  }
```

BC (5.2-2) is included in

```
    if(ir==1){
      u1mrr[1,iz]=4*(u1m[2,iz]-u1m[1,iz])/dr2;}
```

where the indeterminate form at $r = 0$ is regularized as explained for eq. (4.1-1), resulting in the multiplication by 4.

BC (5.2-3) is included in

```
if(ir==nr){
  u1mrr[nr,iz]=2*(u1m[nr-1,iz]-u1m[nr,iz])/dr2;}
```

where the fictitious value u1m[nr+1,iz] is eliminated as explained for eq. (4.1-1), resulting in the multiplication by 2.

The radial group in eq. (5.1-1) at the interior points $r > r_l$, $r < r_u$ is programmed as

```
if((ir>1)&(ir<nr)){
  u1mrr[ir,iz]=(u1m[ir+1,iz]-2*u1m[ir,iz]+u1m[ir-1,iz])/dr2+
              (1/r[ir])*(u1m[ir+1,iz]-u1m[ir-1,iz])/(2*dr);}
```

as explained for eq. (4.1-1).

• The axial derivative

$$\frac{\partial^2 u_{1m}}{\partial z^2}$$

in eq. (5.1-1) is computed as u1mzz with finite differences as explained for eq. (4.1-1).

```
#
# u1mzz, Dirichlet, z=zl, Neumann BC, z=zu
  u1mzz=matrix(0,nrow=nr,ncol=nz);
  for(iz in 1:nz){
  for(ir in 1:nr){
    if(iz==1){
      u1m[ir,1]=u1mb};
    if(iz==nz){
      u1mzz[ir,nz]=2*(u1m[ir,nz-1]-u1m[ir,nz])/dz2;}
    if((iz>1)&(iz<nz)){
      u1mzz[ir,iz]=(u1m[ir,iz+1]-2*u1m[ir,iz]+u1m[ir,iz-1])/dz2;}
  }
  }
```

BC (5.2-4) is included in

```
if(iz==1){
  u1m[ir,1]=u1mb};
```

BC (5.2-5) is included in

```
if(iz==nz){
  u1mzz[ir,nz]=2*(u1m[ir,nz-1]-u1m[ir,nz])/dz2;}
```

where the fictitious value u1m[ir,nz+1] is eliminated as explained for eq. (4.1-1), resulting in the multiplication by 2.

The axial group in eq. (5.1-1) at the interior points $z > z_l$, $z < z_u$ is programmed as

```
if((iz>1)&(iz<nz)){
  u1mzz[ir,iz]=(u1m[ir,iz+1]-2*u1m[ir,iz]+u1m[ir,iz-1])/dz2;}
```

as explained for eq. (4.1-1).

- Analogously, the radial groups

$$\frac{\partial^2 u_{2m}}{\partial r^2} + \frac{1}{r}\frac{\partial u_{2m}}{\partial r}$$

$$\frac{\partial^2 u_1}{\partial r^2} + \frac{1}{r}\frac{\partial u_1}{\partial r}$$

$$\vdots$$

$$\frac{\partial^2 u_6}{\partial r^2} + \frac{1}{r}\frac{\partial u_6}{\partial r}$$

in eqs. (5.1), (5.4) and the axial derivatives

$$\frac{\partial^2 u_{2m}}{\partial z^2}$$

$$\frac{\partial^2 u_1}{\partial z^2}$$

$$\vdots$$

$$\frac{\partial^2 u_6}{\partial z^2}$$

in eqs. (5.1), (5.4) are computed (for u2m, u1, ..., u6).

- Matrices for $\frac{\partial u_{1m}}{\partial t}, \frac{\partial u_{2m}}{\partial t}, \frac{\partial u_1}{\partial t}, \frac{\partial u_2}{\partial t}, \frac{\partial u_3}{\partial t}, \frac{\partial u_4}{\partial t}, \frac{\partial u_5}{\partial t}, \frac{\partial u_6}{\partial t}$ in eqs. (5.1), (5.4) are defined for subsequent display and graphical output.

```
#
# u1mt, u2mt, u1t, u2t, u3t, u4t, u5t, u6t
  u1mt=matrix(0,nrow=nr,ncol=nz);
  u2mt=matrix(0,nrow=nr,ncol=nz);
   u1t=matrix(0,nrow=nr,ncol=nz);
   u2t=matrix(0,nrow=nr,ncol=nz);
   u3t=matrix(0,nrow=nr,ncol=nz);
   u4t=matrix(0,nrow=nr,ncol=nz);
   u5t=matrix(0,nrow=nr,ncol=nz);
   u6t=matrix(0,nrow=nr,ncol=nz);
```

- The programming of the eight PDEs, eqs. (5.1), (5.4), is completed over the 2D domain $r = r_l <= r <= r_u \times z = z_l <= z <= z_u$. First, the Michaelis-Menten functions of eqs. (5.1-3), (5.1-4) are computed.

```
for(iz in 1:nz){
for(ir in 1:nr){
#
#    Michaelis-Menten, probability functions
    q1=pQm1*u1m[ir,iz]/(cm1+u1m[ir,iz])*u2m[ir,iz];
    q2=pQm2*u2m[ir,iz]/(cm2+u2m[ir,iz])*u1m[ir,iz];
    p0=p0m/(1+(g4*u4[ir,iz])^m);
    p1=p1m/(1+(g5*u5[ir,iz])^n);
```

- The derivatives $\frac{\partial u_{1m}}{\partial t}$, $\frac{\partial u_{2m}}{\partial t}$, $\frac{\partial u_1}{\partial t}$, $\frac{\partial u_2}{\partial t}$, $\frac{\partial u_3}{\partial t}$, $\frac{\partial u_4}{\partial t}$, $\frac{\partial u_5}{\partial t}$, $\frac{\partial u_6}{\partial t}$ of eqs. (5.1), (5.4) are computed from the previously computed spatial derivatives in r, z.

```
#
#    PDEs
    u1mt[ir,iz]=D1mr*u1mrr[ir,iz]+D1mz*u1mzz[ir,iz]-q1;
    u2mt[ir,iz]=D2mr*u2mrr[ir,iz]+D2mz*u2mzz[ir,iz]-q2;
     u1t[ir,iz]=D1r*u1rr[ir,iz]+D1z*u1zz[ir,iz]+
                a11*(2*p0-1)*u1[ir,iz]+q1;
     u2t[ir,iz]=D2r*u2rr[ir,iz]+D2z*u2zz[ir,iz]+
                a21*(2*(1-p0)*u1[ir,iz])+
                a22*((2*p1-1)*u2[ir,iz]);
     u3t[ir,iz]=D3r*u3rr[ir,iz]+D3z*u3zz[ir,iz]+
                a31*(2*(1-p1)*u2[ir,iz])-
                a32*u3[ir,iz];
     u4t[ir,iz]=D4r*u4rr[ir,iz]+D4z*u4zz[ir,iz]+
                a41*u1[ir,iz]+a42*u2[ir,iz]+
                a43*u3[ir,iz]-a44*u4[ir,iz]-
                a45*u4[ir,iz]*u6[ir,iz];
     u5t[ir,iz]=D5r*u5rr[ir,iz]+D5z*u5zz[ir,iz]+
                a51*u1[ir,iz]+a52*u2[ir,iz]+
                a53*u3[ir,iz]-a54*u5[ir,iz]-
                a55*u5[ir,iz]*u6[ir,iz];
     u6t[ir,iz]=D6r*u6rr[ir,iz]+D6z*u6zz[ir,iz]-
                a61*u6[ir,iz]-a62*u4[ir,iz]*u6[ir,iz]-
                a63*u5[ir,iz]*u6[ir,iz];
```

- The t derivatives at $z = z_l = 0$ are set to zero to prevent any variation in the dependent variables as defined algebraically by (Dirichlet) boundary conditions (5.2-4), (5.3-4), ..., (5.10-4).

```
    u1mt[ir,1]=0;
    u2mt[ir,1]=0;
     u1t[ir,1]=0;
```

```
        u2t[ir,1]=0;
        u3t[ir,1]=0;
        u4t[ir,1]=0;
        u5t[ir,1]=0;
        u6t[ir,1]=0;
  }
  }
```

The two }s conclude the fors in r, z.

- The eight PDE t derivatives are placed in a single vector ut, to return to lsodes for the next step along the solution.

```
#
# 2D matrices to 1D vector
  ut=rep(0,8*nr*nz)
  for(iz in 1:nz){
  for(ir in 1:nr){
    ut[(iz-1)*nr+ir]         =u1mt[ir,iz];
    ut[(iz-1)*nr+ir+nr*nz]   =u2mt[ir,iz];
    ut[(iz-1)*nr+ir+2*nr*nz] =u1t[ir,iz];
    ut[(iz-1)*nr+ir+3*nr*nz] =u2t[ir,iz];
    ut[(iz-1)*nr+ir+4*nr*nz] =u3t[ir,iz];
    ut[(iz-1)*nr+ir+5*nr*nz] =u4t[ir,iz];
    ut[(iz-1)*nr+ir+6*nr*nz] =u5t[ir,iz];
    ut[(iz-1)*nr+ir+7*nr*nz] =u6t[ir,iz];
  }
  }
```

- The counter for the number of calls to pde2b is incremented and returned to the main program of Listing 5.1 with <<-.

```
#
# Increment calls to pde2b
  ncall <<- ncall+1;
```

- The vector of eight PDE t derivatives is returned to lsodes as a list for the next step along the solution. c is the R vector operator.

```
#
# Return derivative vector
  return(list(c(ut)));
  }
```

The final } concludes pde2b.

This concludes the discussion of pde2b. The numerical and graphical output from the R routines of Listings 5.1, 5.2 is considered next.

5.3.3 Numerical, graphical output

Abbreviated output from the R routines of Listings 5.1, 5.2 follows.

Table 5.3 Numerical output from Listings 5.1, 5.2.

```
[1] 11

[1] 1009

      t        z       u1m(r=0,z,t)
     0.0    0.000        1.000e+00
     0.0    0.250        0.000e+00
     0.0    0.500        0.000e+00
     0.0    0.750        0.000e+00
     0.0    1.000        0.000e+00


      t        z       u2m(r=0,z,t)
     0.0    0.000        1.000e+00
     0.0    0.250        0.000e+00
     0.0    0.500        0.000e+00
     0.0    0.750        0.000e+00
     0.0    1.000        0.000e+00


      t        z       u1(r=0,z,t)
     0.0    0.000        1.000e+00
     0.0    0.250        0.000e+00
     0.0    0.500        0.000e+00
     0.0    0.750        0.000e+00
     0.0    1.000        0.000e+00


      t        z       u2(r=0,z,t)
     0.0    0.000        0.000e+00
     0.0    0.250        0.000e+00
     0.0    0.500        0.000e+00
     0.0    0.750        0.000e+00
     0.0    1.000        0.000e+00


      t        z       u3(r=0,z,t)
     0.0    0.000        0.000e+00
     0.0    0.250        0.000e+00
     0.0    0.500        0.000e+00
     0.0    0.750        0.000e+00
     0.0    1.000        0.000e+00


      t        z       u4(r=0,z,t)
     0.0    0.000        0.000e+00
     0.0    0.250        0.000e+00
```

continued on next page

Table 5.3   (continued)

```
        0.0    0.500    0.000e+00
        0.0    0.750    0.000e+00
        0.0    1.000    0.000e+00


         t       z      u5(r=0,z,t)
        0.0    0.000    0.000e+00
        0.0    0.250    0.000e+00
        0.0    0.500    0.000e+00
        0.0    0.750    0.000e+00
        0.0    1.000    0.000e+00


         t       z      u6(r=0,z,t)
        0.0    0.000    0.000e+00
        0.0    0.250    0.000e+00
        0.0    0.500    0.000e+00
        0.0    0.750    0.000e+00
        0.0    1.000    0.000e+00
                  .        .
                  .        .
                  .        .
         Output for t=0.5 removed
                  .        .
                  .        .
                  .        .

         t       z      u1m(r=0,z,t)
        1.0    0.000    1.000e+00
        1.0    0.250    7.950e-01
        1.0    0.500    6.484e-01
        1.0    0.750    5.596e-01
        1.0    1.000    5.297e-01


         t       z      u2m(r=0,z,t)
        1.0    0.000    1.000e+00
        1.0    0.250    7.950e-01
        1.0    0.500    6.484e-01
        1.0    0.750    5.596e-01
        1.0    1.000    5.297e-01


         t       z      u1(r=0,z,t)
        1.0    0.000    1.000e+00
        1.0    0.250    1.077e+00
        1.0    0.500    9.570e-01
        1.0    0.750    8.090e-01
        1.0    1.000    7.464e-01


         t       z      u2(r=0,z,t)
        1.0    0.000    0.000e+00
```

continued on next page

Table 5.3 (continued)

1.0	0.250	1.300e-02
1.0	0.500	1.848e-02
1.0	0.750	1.647e-02
1.0	1.000	1.470e-02
t	z	u3(r=0,z,t)
1.0	0.000	0.000e+00
1.0	0.250	2.864e-03
1.0	0.500	4.165e-03
1.0	0.750	3.970e-03
1.0	1.000	3.669e-03
t	z	u4(r=0,z,t)
1.0	0.000	0.000e+00
1.0	0.250	2.166e-01
1.0	0.500	2.760e-01
1.0	0.750	2.684e-01
1.0	1.000	2.583e-01
t	z	u5(r=0,z,t)
1.0	0.000	0.000e+00
1.0	0.250	2.166e-01
1.0	0.500	2.760e-01
1.0	0.750	2.684e-01
1.0	1.000	2.583e-01
t	z	u6(r=0,z,t)
1.0	0.000	0.000e+00
1.0	0.250	0.000e+00
1.0	0.500	0.000e+00
1.0	0.750	0.000e+00
1.0	1.000	0.000e+00

ncall =   1218

We can note the following details about this output.

- 11 t output points as the first dimension of the solution matrix out from lsodes as programmed in the main program of Listing 5.1 (with nout=11).
- The solution matrix out returned by lsodes has 1009 elements as a second dimension. The first element is the value of t. Elements 2 to 1009 are $u_{1m}(r, z, t)$ to $u_6(r, z, t)$ from eqs. (5.1), (5.4) (for the spatial grid with $n_r = 6$, $n_z = 21$, $8(n_r)(n_z) = 8(6)(21) = 1008$ points).
- Every fifth value in t and r is displayed, as explained previously.
- ICs (5.2-1) to (5.10-1) are confirmed ($t = 0$).

- BCs (5.2-4) to (5.10-4) are confirmed ($z = z_l = 0$).
- u_6 remains at the IC (5.10-1).
- The computational effort as indicated by `ncall` = 1218 is modest so that `lsodes` computed the solution to eqs. (5.1), (5.4) efficiently.

The graphical output is deferred until the next case (because of u_6 that remains at the IC (5.10-1)).

For the next case, the boundary value of $u_6(r = 0, z = z_l = 0, t)$ is changed from zero (Table 5.3) to one, `u6b=1` in the main program of Listing 5.1. The resulting numerical output follows.

Table 5.4 Numerical output from Listings 5.1, 5.2, u6b=1.

[1] 11

[1] 1009

t	z	u1m(r=0,z,t)
0.0	0.000	1.000e+00
0.0	0.250	0.000e+00
0.0	0.500	0.000e+00
0.0	0.750	0.000e+00
0.0	1.000	0.000e+00

t	z	u2m(r=0,z,t)
0.0	0.000	1.000e+00
0.0	0.250	0.000e+00
0.0	0.500	0.000e+00
0.0	0.750	0.000e+00
0.0	1.000	0.000e+00

t	z	u1(r=0,z,t)
0.0	0.000	1.000e+00
0.0	0.250	0.000e+00
0.0	0.500	0.000e+00
0.0	0.750	0.000e+00
0.0	1.000	0.000e+00

t	z	u2(r=0,z,t)
0.0	0.000	0.000e+00
0.0	0.250	0.000e+00
0.0	0.500	0.000e+00
0.0	0.750	0.000e+00
0.0	1.000	0.000e+00

t	z	u3(r=0,z,t)
0.0	0.000	0.000e+00

continued on next page

Table 5.4 (continued)

t	z	
0.0	0.250	0.000e+00
0.0	0.500	0.000e+00
0.0	0.750	0.000e+00
0.0	1.000	0.000e+00

t	z	u4(r=0,z,t)
0.0	0.000	0.000e+00
0.0	0.250	0.000e+00
0.0	0.500	0.000e+00
0.0	0.750	0.000e+00
0.0	1.000	0.000e+00

t	z	u5(r=0,z,t)
0.0	0.000	0.000e+00
0.0	0.250	0.000e+00
0.0	0.500	0.000e+00
0.0	0.750	0.000e+00
0.0	1.000	0.000e+00

t	z	u6(r=0,z,t)
0.0	0.000	1.000e+00
0.0	0.250	0.000e+00
0.0	0.500	0.000e+00
0.0	0.750	0.000e+00
0.0	1.000	0.000e+00
.	.	
.	.	
.	.	

Output for t=0.5 removed

	.	.
	.	.
	.	.

t	z	u1m(r=0,z,t)
1.0	0.000	1.000e+00
1.0	0.250	7.950e-01
1.0	0.500	6.484e-01
1.0	0.750	5.596e-01
1.0	1.000	5.297e-01

t	z	u2m(r=0,z,t)
1.0	0.000	1.000e+00
1.0	0.250	7.950e-01
1.0	0.500	6.484e-01
1.0	0.750	5.596e-01
1.0	1.000	5.297e-01

t	z	u1(r=0,z,t)
1.0	0.000	1.000e+00

continued on next page

Table 5.4 (continued)

t	z	
1.0	0.250	1.079e+00
1.0	0.500	9.595e-01
1.0	0.750	8.110e-01
1.0	1.000	7.481e-01
t	z	u2(r=0,z,t)
1.0	0.000	0.000e+00
1.0	0.250	1.144e-02
1.0	0.500	1.648e-02
1.0	0.750	1.493e-02
1.0	1.000	1.343e-02
t	z	u3(r=0,z,t)
1.0	0.000	0.000e+00
1.0	0.250	2.553e-03
1.0	0.500	3.744e-03
1.0	0.750	3.609e-03
1.0	1.000	3.355e-03
t	z	u4(r=0,z,t)
1.0	0.000	0.000e+00
1.0	0.250	2.002e-01
1.0	0.500	2.579e-01
1.0	0.750	2.543e-01
1.0	1.000	2.461e-01
t	z	u5(r=0,z,t)
1.0	0.000	0.000e+00
1.0	0.250	2.002e-01
1.0	0.500	2.579e-01
1.0	0.750	2.543e-01
1.0	1.000	2.461e-01
t	z	u6(r=0,z,t)
1.0	0.000	1.000e+00
1.0	0.250	5.552e-01
1.0	0.500	3.058e-01
1.0	0.750	1.828e-01
1.0	1.000	1.458e-01

ncall = 1219

We can note the following details about this output.

- The solution file, out from lsodes, is again 11×1009 (as in Table 5.3).
- The boundary value $u_6(r = 0, z = z_l = 0, t) = 1$ in accordance with u6b=1 (from Listing 5.1)

```
Table 5.3, u6b=0

 t        z          u6(r=0,z,t)
0.0     0.000         0.000e+00


 t        z          u6(r=0,z,t)
1.0     0.000         0.000e+00


Table 5.4, u6b=1

 t        z          u6(r=0,z,t)
0.0     0.000         1.000e+00


 t        z          u6(r=0,z,t)
1.0     0.000         1.000e+00
```

- $u_6(r = 0, z, t)$ does not remain at the IC as in the preceding case (Table 5.3).

```
 t        z          u6(r=0,z,t)
1.0     0.000         1.000e+00
1.0     0.250         5.552e-01
1.0     0.500         3.058e-01
1.0     0.750         1.828e-01
1.0     1.000         1.458e-01
```

- The transit-amplifying (TA) cell density, u_2, remains low.

```
 t        z          u2(r=0,z,t)
1.0     0.000         0.000e+00
1.0     0.250         1.144e-02
1.0     0.500         1.648e-02
1.0     0.750         1.493e-02
1.0     1.000         1.343e-02
```

- The terminally differentiated (TD) cell density, u_3, remains low so that the stem cell production of this final form of tissue is only partly successful.

```
 t        z          u3(r=0,z,t)
1.0     0.000         0.000e+00
1.0     0.250         2.553e-03
1.0     0.500         3.744e-03
1.0     0.750         3.609e-03
1.0     1.000         3.355e-03
```

These results suggest an increase in a_{21} and/or a_{22} (for the production terms in eq. (5.4-2)) and a_{31} (for the production term in eq. (5.4-3)) so that the stem cell differentiation and tissue production are enhanced. This parameter variation is left as an exercise.
• The computational effort remained at `ncall = 1219`.

The graphical output is in Figs. 5.1 to 5.8.

The graphical output for $u_{1m}(r = 0, z, t)$ is in Figs. 5.1-1 and 5.1-2.
The O_2 level increases from a homogeneous IC ($f_{1m}(r, z) = 0$, eq. (5.2-1)) in response to the BC `u1mb=1` ($g_{1m}(r, t) = 1$, eq. (5.2-4)).

The graphical output for $u_{2m}(r = 0, z, t)$ is in Figs. 5.2-1 and 5.2-2.
The nutrient level increases from a homogeneous IC ($f_{2m}(r, z) = 0$, eq. (5.3-1)) in response to the BC `u2mb=1` ($g_{2m}(r, t) = 1$, eq. (5.3-4)).

The graphical output for $u_1(r = 0, z, t)$ is in Figs. 5.3-1 and 5.3-2.
The stem cell level increases from a homogeneous IC ($f_1(z, r) = 0$, eq. (5.5-1)) in response to the BC `u1b=1` ($g_1(r, t) = 1$, eq. (5.5-4)).

The graphical output for $u_2(r = 0, z, t)$ is in Figs. 5.4-1 and 5.4-2.
The transit-amplifying (TA) cell density, u_2, increases from a homogeneous IC ($f_2(z, r) = 0$, eq. (5.6-1)) in response to the increasing $+a_{21}[2(1 - p_0)u_1]$ of eq. (5.4-2), but u_2 remains low (< 0.02).

The graphical output for $u_3(r = 0, z, t)$ is in Figs. 5.5-1 and 5.5-2.
The terminally differentiated (TD) cell density, u_3, increases from a homogeneous IC ($f_3(z, r) = 0$, eq. (5.7-1)) in response to the increasing $+a_{31}[2(1 - p_1)u_2]$ of eq. (5.4-3), but u_3 remains low (< 0.004).

The graphical output for $u_4(r = 0, z, t)$ is in Figs. 5.6-1 and 5.6-2.
The signaling (regulatory) biomolecule 1, u_4, increases from a homogeneous IC ($f_4(z, r) = 0$, eq. (5.8-1)) in response to the increasing $+a_{41}u_1 + a_{42}u_2 + a_{43}u_3$ of eq. (5.4-4).

The graphical output for $u_5(r = 0, z, t)$ is in Figs. 5.7-1 and 5.7-2.
The signaling (regulatory) biomolecule 2, u_5, increases from a homogeneous IC ($f_5(z, r) = 0$, eq. (5.9-1)) in response to the increasing $+a_{51}u_1 + a_{52}u_2 + a_{53}u_3$ of eq. (5.4-5).

The graphical output for $u_6(r = 0, z, t)$ is in Figs. 5.8-1 and 5.8-2.
The signaling (regulatory) biomolecule 3, u_6, increases from a homogeneous IC ($f_6(z, r) = 0$, eq. (5.10-1)) in response to BC (5.10-5), $g_6(r, z = z_l = 0, t) = u6b = 1$.
The contributions of signaling (regulatory) biomolecules 1, 2, 3 are discussed in [1].

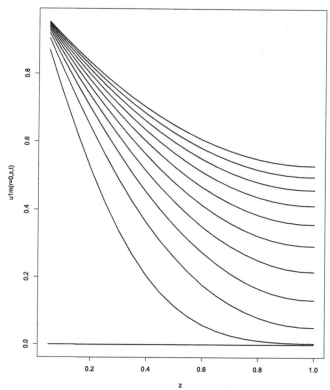

FIGURE 5.1-1 $u_{1m}(r=0,z,t)$ from eqs. (5.1-1), (5.2), 2D, `u1mb=1,u6b=1`.

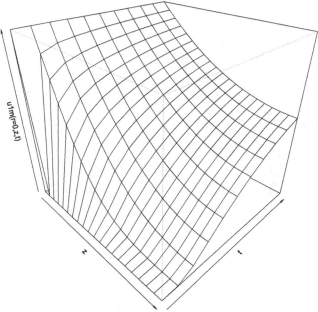

FIGURE 5.1-2 $u_{1m}(r=0,z,t)$ from eqs. (5.1-1), (5.2), 3D, `u1mb=1,u6b=1`.

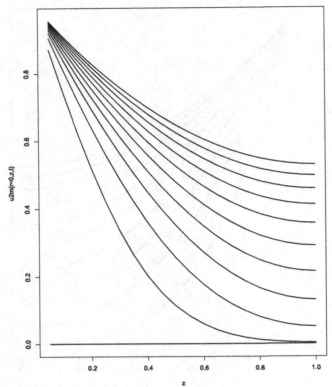

FIGURE 5.2-1 $u_{2m}(r=0, z, t)$ from eqs. (5.1-2), (5.3), 2D, u2mb=1, u6b=1.

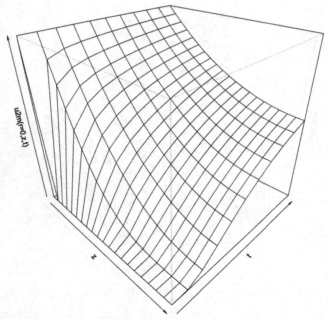

FIGURE 5.2-2 $u_{2m}(r=0, z, t)$ from eqs. (5.1-2), (5.3), 3D, u2mb=1, u6b=1.

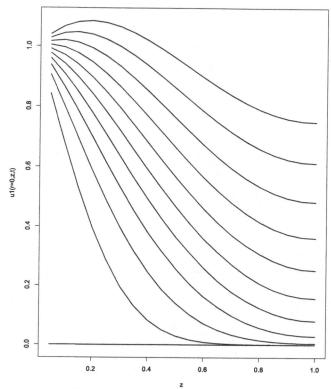

FIGURE 5.3-1 $u_1(r=0, z, t)$ from eqs. (5.4-1), (5.5) 2D, `u1b=1,u6b=1`.

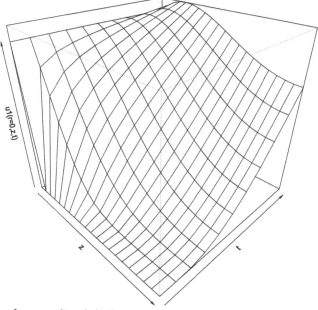

FIGURE 5.3-2 $u_1(r=0, z, t)$ from eqs. (5.4-1), (5.5), 3D, `u1b=1,u6b=1`.

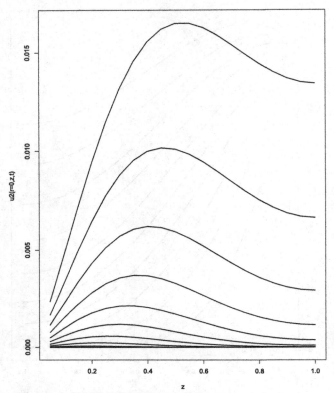

FIGURE 5.4-1 $u_2(r=0, z, t)$ from eqs. (5.4-2), (5.6) 2D, u1b=1, u6b=1.

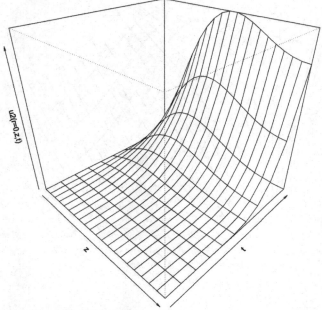

FIGURE 5.4-2 $u_2(r=0, z, t)$ from eqs. (5.4-2), (5.6), 3D, u1b=1, u6b=1.

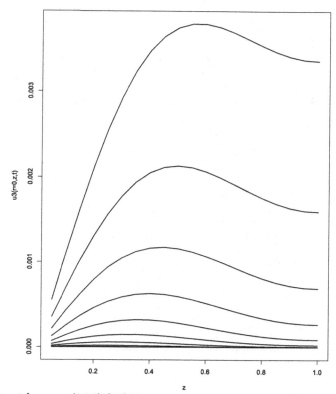

FIGURE 5.5-1 $u_3(r = 0, z, t)$ from eqs. (5.4-3), (5.7) 2D, u1b=1,u6b=1.

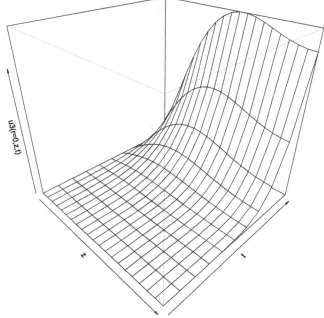

FIGURE 5.5-2 $u_3(r = 0, z, t)$ from eqs. (5.4-3), (5.7), 3D, u1b=1,u6b=1.

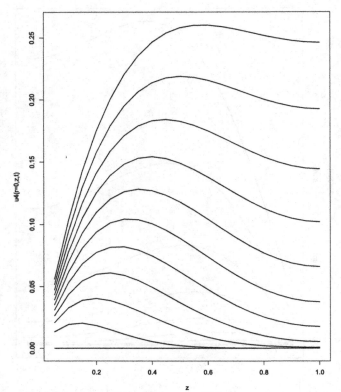

FIGURE 5.6-1 $u_4(r=0,z,t)$ from eqs. (5.4-4), (5.8) 2D, u1b=1,u6b=1.

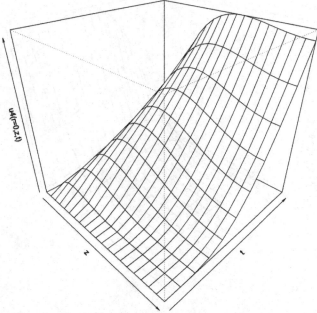

FIGURE 5.6-2 $u_4(r=0,z,t)$ from eqs. (5.4-4), (5.8), 3D, u1b=1,u6b=1.

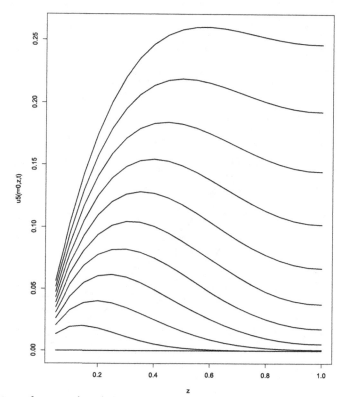

FIGURE 5.7-1 $u_5(r = 0, z, t)$ from eqs. (5.4-5), (5.9) 2D, `u1b=1`,`u6b=1`.

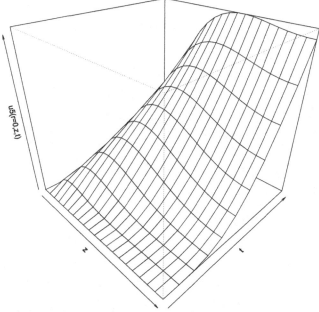

FIGURE 5.7-2 $u_5(r = 0, z, t)$ from eqs. (5.4-5), (5.9), 3D, `u1b=1`,`u6b=1`.

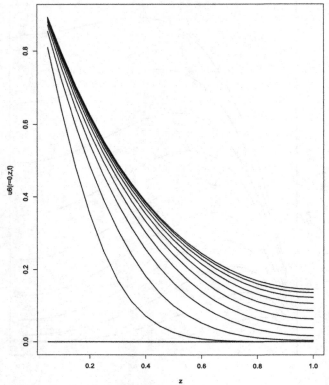

FIGURE 5.8-1 $u_6(r=0,z,t)$ from eqs. (5.4-6), (5.10) 2D, u1b=1,u6b=1.

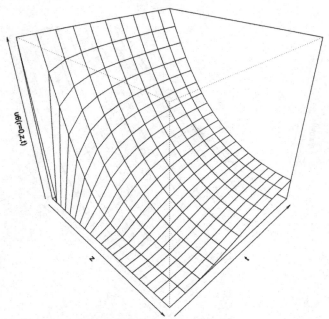

FIGURE 5.8-2 $u_6(r=0,z,t)$ from eqs. (5.4-6), (5.10), 3D, u1b=1,u6b=1.

5.4 Summary and conclusions

In this chapter, six PDEs (dependent variables defined in Table 5.2) are added to the two PDE metabolism model of Chapter 4 (dependent variables defined in Table 5.1) to represent differentiation of the stem cells that are energized by O_2 and a nutrient (metabolism). The terminally differentiated (TD) cell density, $u_3(r, z, t)$, is of particular interest since it represents the synthetic tissue produced by the stem cell differentiation ($u_1(r, z, t) \rightarrow u_2(r, z, t) \rightarrow u_3(r, z, t)$). This synthetic tissue could be the basis for tissue engineering and regenerative medicine.

The eight PDE model (eqs. (5.1), (5.4)) can be used for computer-based experimentation, for example, parameter variation and changes in the model equations or alternate models, to enhance a quantitative understanding of postulated tissue engineering and regenerative medicine.

References

[1] C-S. Chou, et al., Spatial dynamics of multistage cell lineages in tissue stratification, Biophysical Journal 90 (November 2010) 3145–3154.
[2] K. Soetaert, J. Cash, F. Mazzia, Solving Differential Equations in R, Springer-Verlag, Heidelberg, Germany, 2012.

Index

Printed in the United States
by Baker & Taylor Publisher Services